JOCHEN SCHWUCHOW carried ou[illegible] plants between 1986 and 2003 at the universities of Mainz and [illegible] Germany and Ohio State in Columbus, USA. Since 2003 he has worked as a freelance research and teaching consultant at Emerson College in Forest Row, England.

JOHN WILKES originated the Flowform Method in 1970 following periods of work with George Adams and Theodor Schwenk during the 1950s and 60s. After some years further research, project work was begun which led to installations in many countries and the involvement of an increasing number of colleagues. In 2002 an Institute was built at Emerson College providing more adequate space for research into the qualitative effects of rhythmic movement on water.

IAIN TROUSDELL worked with John Wilkes and Nigel Wells from 1976 until the end of 1978 in the early Flowform development phase in England, before returning to New Zealand to pioneer Flowform activities in the Pacific region. He is director of the Healing Water Institute in NZ, and leads a group that is scaling up the supply of Flowform technology internationally.

ENERGIZING WATER

Flowform Technology and the Power of Nature

Jochen Schwuchow, John Wilkes, Iain Trousdell

Sophia Books

Sophia Books
Hillside House, The Square
Forest Row, RH18 5ES

www.rudolfsteinerpress.com

Published by Sophia Books 2010
An imprint of Rudolf Steiner Press

© Healing Water Institute (UK) and Healing Water Institute (NZ), 2010

The moral rights of the authors have been asserted under the Copyright, Designs and Patents Act, 1988

All rights reserved. No part of this publication may be reproduced, stored in a retrieval system, or transmitted, in any form or by any means, electronic, mechanical, photocopying or otherwise, without the prior permission of the publishers

A catalogue record for this book is available from the British Library

ISBN 978 1 85584 240 3

Cover photo by S.E. Gulbekian. Cover layout by Andrew Morgan Design
Typeset by DP Photosetting, Neath, West Glamorgan
Printed and bound in Malta by Gutenberg Press Ltd.

Dedicated to water and its life-supporting capacities

CONTENTS

INTRODUCTION

The noblest of the elements is water.

Pindar (518–438 BC)

This book, a collation of 40 years of research into Flowform®* eco-technology and its effects on water, is published at a time when the poor quality and availability of fresh water worldwide is becoming generally accepted as an urgent global issue whose importance is on a par with global warming.

Global thirst is a vital issue now for about two billion people, who have to struggle daily to find decent water for drinking, cooking and washing.

As long ago as 1970, when John Wilkes invented his Flowform technology, water quality was a major issue in industrialized countries. Since then the added burden on water has continually increased, and humanity's response has as yet fallen far short of what is needed.

The work of the Healing Water Institute is to invent, develop and research technologies based on nature's own perfect methods in order to increase water's capacity to support life. This technology is of course also practically applied and installed, and has already been used in over 2500 projects in some 50 countries.

The other major aspect of our work is to educate as many people as possible about water's fascinating, creative secrets so that we all understand the vital nature of this wonderful element. Unless we profoundly appreciate it, how can we be moved to change things to help it? When we help water we are helping ourselves, and all of nature.

The Flowform work has been carried out by a relatively small number of colleagues around the world who have seen the practical value of creating designs that re-energize human-captured water with heart rhythms and the dynamics of mountain cascades, before putting it back to work in our industrial, agricultural and domestic situations.

Presently humanity captures drinkable water from 65% of the world's natural fresh-water sources. UNESCO tells us that within 35 years this will rise to 90%. So what does this mean?

With our deplorable record of polluting this water, chemically, organically and energetically, it means that the world's water will no longer be able to support all life on this planet, as it has done previously through aeons.

We pollute the seas through discharge from our farms, factories and not least from our homes. It is of vital importance that we change the way we think, for our polluting practices all stem from the way we work, or don't work, with nature. And

* Flowform is a registered trademark.

these practices all originate from our thoughts, and lack of understanding.

The Flowform invention is a prime example of a new way of thinking, a new paradigm of inventive, 'living thinking', where practical solutions are developed by studying nature's subtle metamorphic and rhythmic processes. Such study has led to this effective Flowform method of water treatment – which is actually nature's own method. To be more accurate, the work combines two of nature's best-known rejuvenation methods – the active vitality of the mountain stream and the rhythmic pulse of nature in all living things.

This aspect of the 'living pulse' may seem somewhat non-scientific to some, but again this is a paradigm issue, and thus one of human thinking. Rhythm exists wherever there is life, even when there is no pulsing heart. Water is the carrier of these subtle and essential rhythms, mediating the creative forces continually at work throughout nature. Life without rhythm is actually death.

When water is removed from its natural context and treated in a purely mechanical and chemical manner, its capacity to support life is reduced. Nature has known what it is doing, so to speak, and unless we learn quickly and profoundly from it we will fail to be in harmony with it. And nature, as we are discovering, is far stronger than humanity wherever disharmony arises.

Flowform technology offers the chance to return captured and spoiled water to its natural context, invoking nature's own wisdom even in an agricultural, industrial and domestic environment.

Much more research is needed into this whole area, and recent decades have only seen the start of what might develop.

Since the turn of the new century we find that our work with rhythmic and surface influences on water is meeting with greater understanding and that new, exciting opportunities are developing. University water science of the last 15 years, influenced by quantum physics, has also entered the same theoretical fields that we have been exploring for decades. Based on Rudolf Steiner's scientific suggestions in the 1920s, the present Healing Water Institute's work draws on investigations by highly qualified scientists such as Ehrenfried Pfeiffer, Theodor Schwenk, Eugen and Lily Kolisko, and George Adams. John Wilkes studied under the latter and became assistant in his water movement research.

John Wilkes, Nick Thomas and Nigel Wells first established the Healing Water Institute in 1975. It is now registered as a Charitable Trust (no. 1133741), and is based at Emerson Village, Sussex, England, with sister institutes in New Zealand and the USA, and other scientific associations developing elsewhere.

The concept of energetic water quality introduced in this book inevitably gives rise also to that of energetic water pollution. This is a new idea to many western thinkers, though not to Asian and other traditional thinkers who are aware of related concepts such as that of 'prana' or 'chi', the basis of longstanding and effective Vedic and Taoist sciences.

Modern western quantum physics has advanced the concept of energy information emitted as signature frequencies from the material molecular level. Water is vul-

John Wilkes (right) and Costantino Giorgetti (Trustees) in front of the Healing Water Institute building at Emerson Village, Sussex, England

nerable to and creative with this energetic information to a remarkable degree. Throughout evolution, of course, nature has managed to renew water so that the creative information it contains can sustain living things.

The interference of human inventions, based on mechanistic thinking sundered from nature, too often works counter to this ongoing, natural creative energy that is central to life.

We ignore the configuring forces at work in nature at our peril. As part of this urgently needed new paradigm of living thinking in tune with nature we need to consider the source of creative life processes as supersensible and acting from invisible realms into the physical world, rather than as some biochemical coincidence that has evolved accidentally. This is a small theoretical step to take, bearing in mind that modern physics has for decades been developing concepts far removed from human sensory perception.

This book therefore presents the pioneering efforts by a team of highly qualified and committed researchers and designers inspired by the work of John Wilkes and George Adams. They have sought to discover more about the influences of formative forces* on water and living things.

Our key question remains: 'How can mankind effectively reintroduce nature's rhythmic enlivening processes into water in order to enhance its life-supporting qualities?'

This book is published with sincere gratitude to all those who have dedicated some portion of their lives to the Healing Water endeavour, through their interest in rhythm research, the influence of formative forces on water, and via water upon nature.

Bringing together a collation of many decades of research is a team effort and the authors especially want to thank publisher Sevak Gulbekian and the editorial team of Matthew Barton and Eileen Lloyd as well as Alison Trousdell and Constantino Giorgetti for their patient and determined help.

All who wish to help us in this urgent task of improving water's quality can contact us via the following websites:
www.healing-water.org
www.flowform.net

February 2010

* 'Formative forces' are those supersensible forces which work into the physical world to fashion the forms of living organisms.

1. WATER QUALITIES SUPPORTING LIFE
Background Concepts

Water is the principle of all things. Everything originates from water, and into water all things shall return again.

Thales of Milet (AD 625–545)

Traditional approaches to water's energetic qualities

Since the beginning of recorded history, humans established a strong relationship to water as a life-sustaining substance. In a wide variety of cultures and civilizations stretching over several thousand years, water held a unique place in the physical, intellectual and spiritual lives of people. And some basic insights into water appear to transcend both culture and time.

For ancient people, collecting water often took great effort and labour, and they were aware of its indispensable nature and energetic qualities. Rather than an inanimate liquid, water was considered to be an essential and primordial substance from which the universe arose and was made manifest. People's inner relationship to water was often one of ritual reverence and worship. Water was considered to be a conscious and life-giving entity filled with spiritual and divine beings, and a link between the physical and spiritual world.

Reverence for this life-giving substance was frequently expressed at places where it sprang from the body of the earth to its surface. These locations included springs, streams, rivers and lakes, which became sacred sources of nourishment, and the sites of offerings, rituals and prayer.

This kind of water worship was practised, evidently, in many of the major civilizations of the world, including those of Sumeria, Babylon, Egypt, Troy, Greece and Rome, as well as in indigenous cultures of the American Indians, Australian Aborigines, and some of the African tribes (Marrin 2002). Their rituals were intended to show reverence and give thanks for an essence which they experienced as life-giving and with which they felt a strong inner connection.

Many ancient cultures around the world recognized that the interaction between water and fire, represented by the sun, is fundamental to all forms of life on earth. Frequently, according to their mythologies, creation was preceded by an original state of chaos, which was identified as a void, abyss or primordial sea (Marrin 2002).

In ***Sumerian*** history, the god of the watery underworld Enki embodied both water and wisdom, and was believed to have evolved from the same underworld ocean that created heaven (Marrin 2002). Throughout Sumerian history, water was associated with heaven and worshipped as a substance of wisdom and magic.

In ***Greek*** mythology, the River Styx formed the boundary between earth and the underworld or Hades, and was said to circle Hades nine times. The Styx served as a crossroads where the worlds of the living and dead and of mortal and immortal encountered one another.

In ***Egyptian*** mythology, water in the form of moisture was worshipped in the god Tefnut, who was symbolized by the moon. The Egyptians, like the Sumerians and Babylonians, believed that the heavens and the earth and all its life were created from the celestial or primordial waters.

In ***Mesopotamian*** mythology, the original waters were associated with the creative 'chaos' that existed before the world was made (Marrin 2002).

In the **Romantic Age** in the West (18 c.) this intuitive understanding could still be found, with water considered to be the carrier of formative processes. In Romantic nature-philosophy, it was seen as the substance archetypal of a formless quality that was both open and highly receptive at the same time – which Novalis in his *Fragments* called 'sensitive chaos'.

In the ***Polynesian South Pacific*** the Maori people considered water (*wai*) to be purely spiritual, created continually anew as rain in the form of tears falling from the father sky god Rangi yearning for his wife Papa, the earth. The Maori experienced and treated water by sensing its spiritual energy quality (mauri) and considered themselves a portion of water itself. Types of water were *waiora* (purest, refreshed holy water), *waimaori* (running freely), *waikino* (polluted or spoilt) and *waimate* (dead water, with no life force). Sea water also had gradations with *waitai* being enlivened surf water. There were specific etiquettes of behaviour and inner response required for each of these types of water (I. Trousdell personal communication).

Throughout Asia, both Taoist and Vedic sciences perceived 'prana' and 'chi' as the essence of water, through which life imparts form to all things. This life essence could be worked with consciously.

Dowsing is an art and technique of water sensitivity that, historically, humans developed at an early stage, allowing them to expand their abilities beyond three-dimensional and sensory limitations. Used mostly in the past to find water, the art of dowsing – also called radiaesthesia – has now been extended to encompass many aspects of human life, such as medical diagnosis.

The earliest sign of dowsing usage comes from an inscription on a grave in Brittany that is 4500–5000 years old (J. Rasmus 1998). The fact that dowsing has existed in various forms for several thousand years highlights ancient wisdom about water's energetic qualities.

Dowsers claim that they can perceive these energetic qualities (frequently called E-rays or earth-rays) with the help of rods of special shapes. Traditionally, the most commonly used rod was a Y-shaped branch from a tree or bush (often hazel twigs were chosen), and many dowsers today use a pair of L-shaped metal rods to detect water, metals or general objects.

Some authors who have done research into the field of dowsing, suggest that these abilities may be explained by postulating human sensitivity to small magnetic-field gradient changes (Baker 1983, Presti and Pettgrew 1980, Rocard 1981).

The persistence of dowsing through the centuries can be regarded as a human quest for understanding the qualities of water at a deeper level beyond ordinary sensory perception.

Thus knowledge of certain energetic qualities of water beyond mere chemical and physical properties was inherent in many traditional cultures. By contrast, in the modern world we treat water as a mere commodity to irrigate crops, produce hydroelectric energy, provide transportation, and as a coolant or solvent.

Recently science has discovered that water is not only fundamental to the energetic processes sustaining our biosphere and to supporting and mediating biological processes, but also plays an essential role in star formation from interstellar gas and dust clouds. Now, in the twenty-first century, we face decisions of immense importance regarding water that will affect our well-being and the future quality of our lives, as well as all living organisms on our planet. Instead of treating water as a mere commodity to meet our immediate needs, we need a wider context of insight as the basis for decisions about our impending – and many say already present – water crisis.

Water as an omnipresent and life-giving substance

Water is the most common substance on the earth's surface and as such carries and supports every aspect of life. It accounts for about 95% of the fertilized human egg, and the amount of water in a mature human body is more than 70%. About the same amount (71%) of the earth's surface is covered with water. Of all the water on earth, 97.4% consists of salty sea water, 2% is frozen ice, and only 0.6% of the earth's water reserve is fresh, potable water in a continuous cycle of evaporation, precipitation, flow and percolation.

Water moves in intimate relationships to the surfaces over which it flows. Over longer periods it erodes and shapes them in accordance with its own rhythmic nature, as in riverbeds, beaches or rocks the world over. It will form surfaces, such as when it moves through the air as clouds or when it is contained in some way. Water also has the ability to form internal surfaces or layer sheaths within itself.

The quality of water is difficult to define and does not solely depend on its chemical composition. One aspect concerns the relation of water to its ecological context and to its capacity to support life. Another aspect concerns its purity. Unwanted chemical and biological constituents create pollution while, on the other hand, the lack of certain trace elements reduces the life-supporting quality of water. In its ecological context, water used to possess the correct trace elements in the correct proportions, introduced into it quite naturally.

Even after chemical and biological purification, water contains certain electromagnetic frequencies that can be supportive or harmful. Water quality therefore has at least three major aspects, all of which we need to take into account:

1. **Chemical**: basic inorganic element combinations within water.

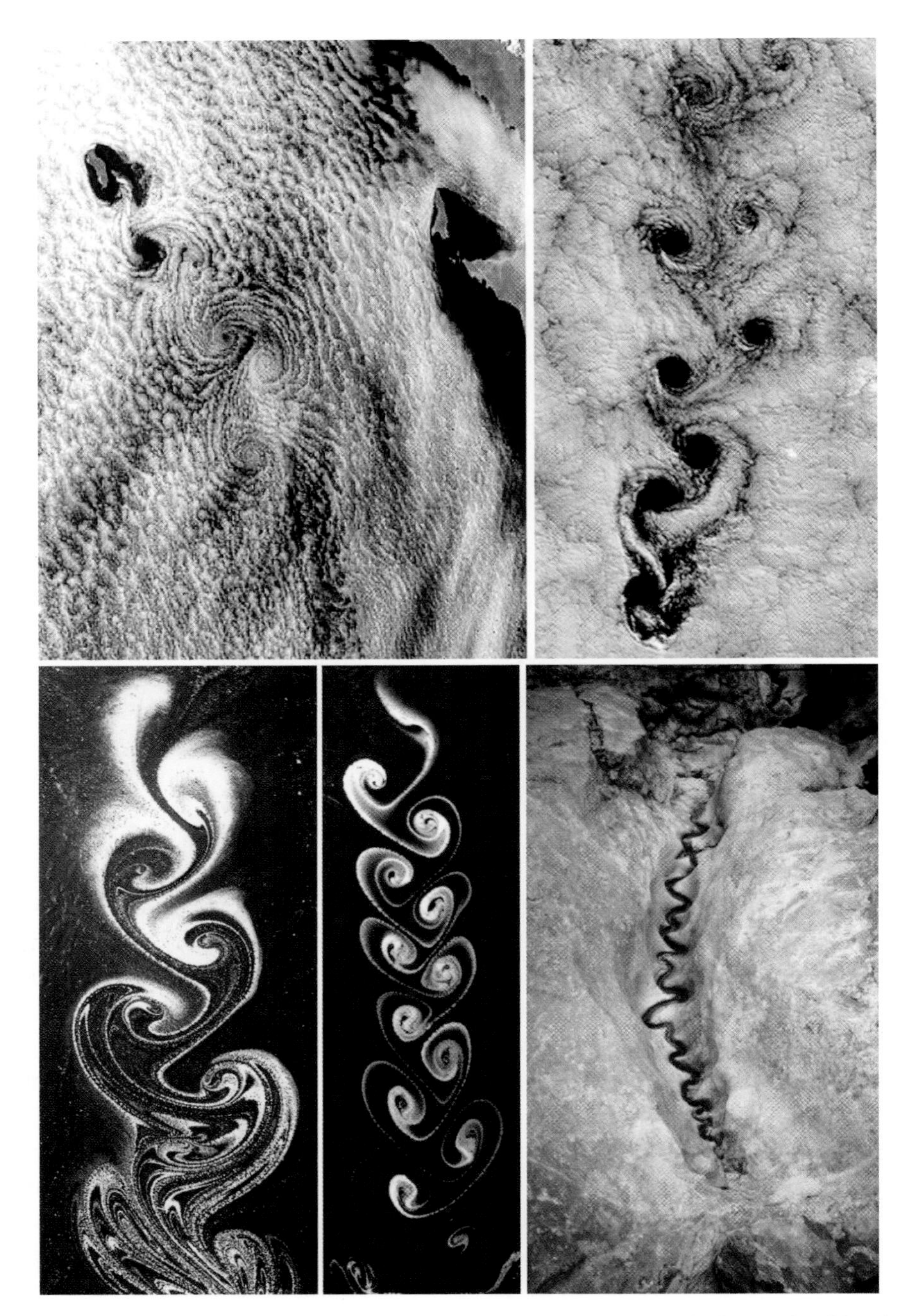

Fig. 1 *Path of vortices in cloud formations* (top), sometimes generated when a cloud bank passes a high mountain point. *Path of vortices in water* (bottom left and centre), made by a straight-line movement through a solution of glycerine and water dusted with a fine powder to render visible. *Meander formation* in rock (bottom right)

2. **Organic**: micro-organisms, algae and other materials, dead or alive.
3. **Energetic**: condition of freshness, or the internal microstructure condition of the water.

Energetic water quality can be assessed through the various frequencies emitted from it, as well as through other methods such as scientific crystallization, showing levels of entropy (disorder and order) within the internal harmonic structures.

Only in the absence of harmful chemical, organic *and* energetic influences can water be considered as life-supporting for the organism and in balance with the environment.

Given that this balance has been impaired, how can life processes still be influenced in a positive way, and water's mediating capacity be enhanced?

To restore water to this original qualitative state is not simply a matter of chemical and physical manipulation. Such action needs to be supported with specific treatments that affect water's energetic qualities, such as potentizing (used in homoeopathy) and rhythmical movement.

Water enables the individual organism to be rightly embedded within its immediate and also cosmic environment. We are increasingly eroding this capacity as we continue to pollute water and use it technologically without being aware of its life-supporting function. We alter or destroy water's complex structures by passing it through pipes, turbines or artificially straightened rivers. As we use water for transportation and industry, for effluent treatment and energy production, it is increasingly cut off from its natural state.

Artificial urban landscapes lead to significant changes in the natural water cycle. In a natural forest, more than two thirds of the rain water evaporates, about a quarter filters into the ground, and less than 5% of all water runs off. In contrast, in a typical urban region only one third of the water evaporates, about a tenth filters into the ground, and more than half of all rain water runs off, leading to a lowering of the ground water level, and – when combined with the influence of straightened rivers – creates severe problems of flooding (Geiger and Dreiseitl 1995). As vital and pollutant-free natural water supplies become more scarce, we need to support and enhance water's natural, life-giving function, finding ways to restore it to its original vigour and vitality by nurturing it on its own specific terms.

Water – a substance with surprising properties

For a long time, H_2O appeared to be the pivotal centre of chemical knowledge, and water one of the best-known substances. All atomic weights were calculated with respect to hydrogen and oxygen. Harold Urey's discovery in 1931 of deuterium oxide (D_2O), or heavy water, put paid to that illusion and shook the foundations of chemistry at the time. It demonstrated that despite water's omnipresence and significance for all organisms it was not well known at all. Urey found a method of concentrating deuterium – a type of hydrogen containing one proton and one neutron in its nucleus – by distillation of liquid hydrogen. Thus it was proved that there was not only one element of hydrogen but it existed in a mixture of different types of hydrogen, called isotopes, with different nuclei and varying atomic weights.

From the point of view of physics, water also has very anomalous and surprising properties:

- Compared with similar chemical substances, according to its molecular weight the freezing point would be expected to be around −120°C, and the boiling point around −75°C; therefore water would not exist in the liquid state under normal temperature conditions. However, water is prevented from evaporating at these low temperatures, because the water molecules, due to their dipole charges, combine to form long chains with hydrogen bonds. At our body temperature of 37°C, approximately 300 to 400 molecules combine to form clusters. The colder the water is, the longer these chains normally become (Ludwig 1991).
- Upon freezing, due to its expansive crystal structure, water increases in volume, instead of shrinking like most other substances.
- Water's highest density is at +4°C (anomaly point). Below this temperature it expands instead of contracting, causing ice to float on top of warmer water. This is the reason why lakes and oceans do not freeze at the bottom first (Schiff 1995).
- Water needs twice as long as one would expect, by comparison with other chemical substances, to absorb or give off heat.
- Water requires much more thermal energy for evaporation than comparable substances. If this were not the case, large volumes of the oceans would evaporate every day and return to earth at night (Kröplin 2001).
- Water has a much higher surface tension than comparable substances, in an order of 70 × 10(-3) N/m instead of an expected 7 × 10(-3) N/m. (Newton per metre is the unit of force acting on the peripheral surface, which tries to reduce the surface area.)
- Water is one of the best solvents, since 84 out of the known 103 elements of the earth are soluble in water. Therefore water is essential for the transport of nutrients in all organisms (Kröplin 2001).
- Another anomaly of biological importance is the abnormally high dielectric constant of water, which permits very large electric fields within living cells (Schiff 1995).
- The critical temperature is at 374.2°C instead of 50°C. This refers to the temperature above which a gas can no longer be liquefied, however high the pressure

In fact, it is precisely these anomalous properties of water that have made life possible on earth (Schiff 1995). For example, the fact that water drops are stable due to the high surface tension of water is important for sustaining life. When water is carried through the xylem up plant stems, the strong intermolecular attractions hold the water column together and prevent tension rupture caused by transpiration pull. Other liquids with lower surface tension would have a higher tendency to 'rip', forming a vacuum or air pockets and rendering the xylem water-transport inoperative.

However, these anomalous properties cannot account for the capacities of cluster formation, communication or memory of water that have been studied in some detail by Jacques Benveniste, Masaru Emoto, Del Giudice and Preparata, and others.

Information transport and water memory

In recent years, the so-called 'memory of water' has been the subject of controversial discussions. In several independent laboratories around the world, indications have been found that water is capable of storing and transmitting information about substances that have been in contact with it (Davenas et al. 1988, Belon et al. 1999, 2004, Ludwig 1991. For further references see Schiff 1995).

French immunologist Jacques Benveniste announced that a homoeopathically diluted solution of antibodies could activate white blood cells (basophils) without relying on a chemical reaction, indicating a capacity of water to retain a memory of substances that it no longer contains (Davenas et al. 1988). In 2004, Madeleine Ennis and colleagues published an article in *Inflammation Research* claiming: '... it has been shown that high dilutions of histamine may indeed exert an effect on basophil activity ... We are however unable to explain our findings and are reporting them to encourage others to investigate this phenomenon' (Belon et al. 2004). Following up on a study they had published in the same journal in 1999, the researchers concluded that a memory effect did exist (Belon et al. 1999).

Michel Schiff (after excluding all publications in homoeopathic journals) reported that by 1994 he had already found 25 scientific articles published by 17 different research groups reporting effects at high dilution (Schiff 1995).

Benveniste also conducted numerous transmission experiments, in which chemical information seemed to be transmitted to a solution through an electronic device without the concomitant transport of molecules (Schiff 1995).

Masaru Emoto uses water crystals as an indicator of the quality of drinking water. He also claims that crystal formation can be influenced by words, names, conscious thought or music. Emoto's findings support the idea that water reflects human consciousness, and that it is capable of storing and transmitting information. He concludes that our knowledge of water is still extremely poor and limited (Emoto 1999). However, the scientific relevance of Emoto's work needs to be questioned since only one picture of each sample is ever shown in his publications. He does not appear to mention his criteria for choosing an individual image from the series of a hundred or more pictures he claims to take for each sample.

Hydrogen bonding and the bipolar nature of water

Water differs from other liquids in that it forms a 'two-phase system'. This means that in addition to a disorganized proportion of water molecules it also appears to have a highly ordered proportion. The ordered phase is also referred to as crystalline fluid, since it has the same high degree of order as a crystal.

While the energetic bonds in the normal, disordered fluid phase are rather low, the inter-molecular forces in the crystalline-fluid part represent a very high level of energy, leading to connections between several hundred water molecules. These inter-molecular bonds are referred to as hydrogen bridges. Several such bonds produce a complex lattice system with an unmeasurably high number of possible structures for the crystalline-fluid phase of the water. This lattice system can vibrate like an instrument and reveal a large number of its own individual frequencies.

Water is a polar molecule: it has separate positive and negative charges and thus exists as an electric dipole. This is due to the 104.5° angle of the hydrogen bonds to the oxygen atom (Block 2004). The oxygen atom attracts the electron of the hydrogen atom, thus the region around the oxygen is negative compared to the positive region around the hydrogen atoms. Because of this molecular configuration, water molecules attract one another and are linked by these hydrogen bonds, forming so-called clusters. This clustering imparts a crystalline property to the water (structural water). Water has the capacity to align into 400–500 hydration layers (Pollack 2001). At body temperature, there are about 300–400 water molecules cross-linked into a cluster. In living organisms, the clusters form hydration layers around biological molecules.

Due to their bipolar nature, water molecules line up in an electric field. If the field direction is reversed, the molecules will about-face. As long as the frequency of the imposed field is not too high, water molecules will continue to flip with the imposed frequency. When the frequency is raised beyond a critical value, the water molecules will no longer be able to respond in timely fashion. For ordinary water, the critical frequency for this weakening is 20 GHz. In structural water, the critical frequency drops to 10 kHz. Frequencies below these limits allow the structural water to move in resonance with the entraining or influencing frequency. Alternating current frequencies (50 Hz, 60 Hz) are well within this range and are known to deleteriously affect many biological processes (Block 2004).

Cluster formation, coherent domains and electromagnetic information exchange in water

Each molecule has an oscillatory pattern (resonance frequency) that can be determined by spectroscopy. It is known, through spectrographic analysis, that water and other dipole molecules can be entrained or respond to exogenous oscillatory patterns by rearranging their cluster patterns.

This frequency spectrum in water is thus a physical manifestation of its geometric structure. It changes in a specific way over the course of various vital processes, and characteristics of these vital processes can be read off from the water spectroscopically. Because of these findings it has been suggested that drinking water should be subjected to a detailed spectroscopic analysis, capable of measuring all frequencies from zero up into the megahertz range (Gross 2000b). Peter Gross supports the idea that in future we will be able to define water quality not only chemically but also above all spectroscopically (Gross 2000b).

The two Italian physicists Del Giudice and Preparata have shown that closely packed atoms or molecules have a collective behaviour and, by forming higher order clusters or 'coherent domains' containing millions of molecules, act as a whole rather than as isolated particles (Del Giudice and Preparata 1994, Del Giudice et al. 1988). These clusters form structures similar to crystals, with the crystalline lattices vibrating at high frequencies. According to their theory, molecules spontaneously go from a chaotic to this coherent, ordered state if the ordered state has a lower energy level (Schiff 1995).

In isolated atoms or molecules, only the short-range electrostatic forces (hydrogen bridges) are relevant, which become negligible beyond the nearest neighbours. According to quantum mechanics, once atoms or molecules are sufficiently tightly packed (as they are in water), the long-range forces start to play an important role (Schiff 1995). These long-range electromagnetic fields, which modify the energy balance and lead to qualitatively new effects, can be transmitted by coherent domains present in water (Del Giudice and Preparata 1994). According to Preparata and Del Giudice, the stability of a substance can only be achieved through the existence of these long-range forces that lead to collective behaviour of a higher order. These forces can also account for the existence of liquid water: they make it energetically more attractive to molecules to be in the condensed form of a liquid than in the more dispersed form of a vapour (Schiff 1995).

Therefore clusters appear in water on at least two fundamentally different levels: lower order clustering appears at the level of hydrogen bonding, and higher order clustering at the level of these structured coherent domains, which have laser-like properties (Del Giudice et al. 1988).

This higher-order organization of water molecules could provide the physical basis for the apparent ability of water to 'remember' previous contact with other substances (Schiff 1995).

Water clusters apparently send out typical energy signals which depend on the movement of their individual molecules. When these signals are recorded, the picture resembles a relief map (Gross 2000b).

Jürgen Schulte of the University of Michigan suspects that homoeopathic information is stored in these clusters (Gross 2000b). This could be the reason why homoeopathic substances still work at concentrations that are so highly diluted that not even a single molecule of the original substance is still present in the solution.

Electromagnetic pollution and water purification

According to the German water researcher and physicist Wolfgang Ludwig, water can store the information that has been imprinted on it at the level of certain frequencies and transmit such information to other systems, such as living organisms (Ludwig 1991). So water may not only contain positive and healing signals, but also all information on harmful materials with which it has come into contact.

As soon as the water comes into contact with harmful substances such as lead, cadmium or similar materials, the resonance properties of the water change, creating a completely new spectrum since each harmful substance has its own range of frequencies. These seem to be transferred to the water as soon as the water comes into contact with the material in question. The photon (quantum light) spectrum of contaminated water thus differs very significantly from that of clean water: the latter demonstrates an abundant photon exchange with the environment and with the organic cells contained in the water.

Water clusters have been implicated in the atmospheric processes of acid rain formation and the anomalous absorption of radiation in clouds, although the processes are not well understood (Paul et al. 1997).

Only few scientists so far acknowledge the energetic and informational problems of pollutants in drinking water. The water researchers Wolfgang Ludwig and Peter Gross conclude that water, even after treatment and purification, contains certain electrical frequencies which, depending upon their wavelength, can be destructive and/or injurious to our health (Gross 2000b, Ludwig 1991). By further analysis, these frequencies can be tracked precisely to those detrimental substances that were detected in the water before treatment.

These frequencies can be picked up in a similar way to radio waves, as is already being shown by a research group of chemists at the University of California in Berkeley with an infrared absorption spectrometer (Paul et al. 1997).

This transfer of electromagnetic vibration has also been confirmed by Wolfgang Ludwig in a trial: from a sealed ampoule floating in the water, containing two electrodes, the frequency of the homoeopathic solution in the ampoule was reported to be transmitted to the water – right through the completely sealed ampoule (Gross 2000b).

Water contaminated with information can be dealt with in various ways. In order to delete all the information stored in water, energy must be applied. Almost all of this information disappears if water is heated to 400°C, or if it is subjected to vigorous swirling (Gross 2000a). Several developments, such as vigorous vortical lemniscatory movements generated in Flowform cascades or the vigorous shaking process in homoeopathy, make use of the principle of the swirling of water. As a natural process this is known from streams and rivers: water is swirled around in the curves and bends, therefore it is subject to a self-cleaning process, both chemically and physically.

The naturally swirling movement appears to be essential for the enhancement of vital qualities in water. Measurements have confirmed that the harmful information stored in certain frequencies can be deleted by multiple swirling. If the water however is still chemically con-

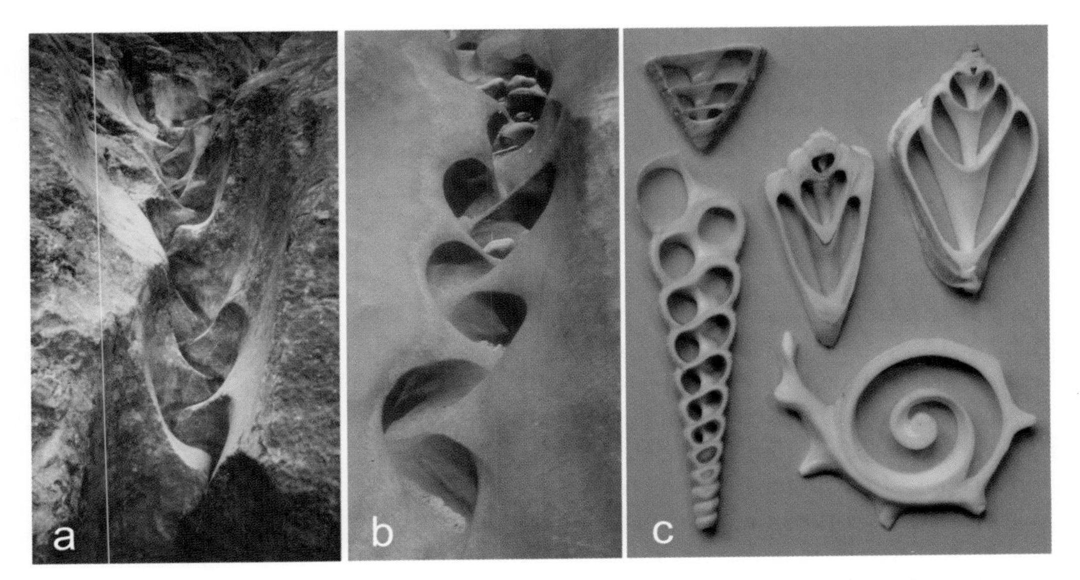

Fig. 2 *(a, b) Rock meanders* showing a 'path of vortices' sequence, formed through erosion processes. *(c) Cross sections of different shells* showing vortical and spiral patterns evocative of vortices in water

taminated, the harmful frequency will soon reoccur, at least in normal tap water (Gross 2000b).

Many researchers are convinced that water has a sufficiently large supply of different cluster structures to fill whole libraries with information. Due to the complexity of the processes going on in water, even the most powerful computers cannot reproduce the dynamics of the signal exchange that takes place in large molecule systems. As a result, the consequences of cluster formation still remain largely unexplained.

All biological interactions occur in water since, on average, there are ten thousand molecules of water per molecule of protein. If every substance that comes into contact with water leaves an electromagnetic imprint in it, then, taking into account the high quantity of water contained in all living organisms, it becomes hard to imagine the volume of information processed and stored in the watery medium of each living cell, let alone in each one of us.

Chemical and electromagnetic models for biological systems

Commonly, molecules are described as encountering each other by chance and fitting into an interaction of the 'lock and key' model. Even with biological arrays (arrangements of interrelated factors), this model provides a very low probability of meeting, and requires a very long time to happen. In a more recent model, molecules interact by co-resonance and need not actually touch as long as they are within an energetic field radius. A series of articles have been published concerning this energetic model of biological reactions and regulation of the interactions between a signal and receptor molecule (Benveniste 1998, Smith 1987, Szent-Gyorgyi 1988).

In living systems, relatively long-range electromagnetic fields engage in resonance matching and coherent amplification between distant molecules. This occurs as long as emission and absorption spectra match. Thus, non-resonating, unwanted signals are excluded. Jacques Benveniste found evidence that this electromagnetic model can account for intermolecular interactions. In experiments over many years, he has recorded the resonance frequencies of signalling molecules, matched them to the harmonics in the audible range, digitized them and then, using a computer sound card, played the recording to the receptor molecules, evoking an appropriate response in the specific receptor molecule, just as if the molecules were in local contact. Benveniste has suggested that the effects of specific biological molecules (histamine, caffeine, adrenalin, insulin), as well as viruses and bacteria, are due more to electromagnetic interaction than direct contact.

Water's capacity to store and process information at a different level has been examined and used in micro-fluidics which, since its emergence in the 1990s, has become a powerful tool for a wide variety of applications in biotechnology, chemistry, physics and engineering. By studying processes in channels with typical dimensions of tens to hundreds of micrometres, researchers can conduct controlled reactions while economizing on the consumption of possibly scarce materials. A micro-fluidic channel carries tiny water droplets (volume $\sim$250 pl) in a continuous stream of oil. The droplets act as micro-reactors in which the reagents are rapidly mixed and are then transported with no dispersion (Epstein 2007). Fuerstman et al. (2007) and Prakash and Gershenfeld (2007) report their use of micro-fluidic technology to construct streams of droplets (liquid-in-liquid) and bubbles (gas-in-liquid) that can encode and decode information or perform logical operations.

Rhythmic patterns in nature

Wherever we look in nature, whether at the patterns and movements of water or air, the structures of minerals and crystals, or growth processes in plants, we encounter wavelike patterns and rhythmical formative processes with their polar opposites of contraction and expansion (Kröplin 2001).

Patterns arising in streaming water and air are often surprisingly similar to patterns that can be found elsewhere in living nature. For example, as described in the book *Sensitive Chaos* by Theodor Schwenk, many unicellular water animals, plants, snails and shells, animal horns and bones have incorporated the archetypal spiralling and vortical movement of water into their shapes (Schwenk 1996, also see Figs 2, 3, 4 and 7).

Schwenk showed how a gentle thrust of (coloured) water released into still water (depending on direction, intensity and proportion) can take on the form of a pulsating jellyfish medusa, an unfurling fern, a human larynx, a cochlea or bone formations. Once the movement comes to a standstill, the form collapses, showing therefore how strongly movement is related to dynamic formative processes. Depending upon the particular conditions, obstacles placed in flowing water create vortices that demonstrate a similarity in structure to formations in clouds or in the bark and knots of a mountain oak, olive or cypress tree (Schwenk 1996, and see Figs 1 and 11).

D'Arcy Thompson suggested that many organic forms show symmetries corresponding closely to those of vibrating bodies of similar shape, and many others are strikingly reminiscent of shapes exhibited by vortices and splashes, such as drops falling into a liquid. He also studied physical proportions of different animal species such as crabs and fish, and found geometrical transformations that can

Fig. 3 *Spiral forms:* (*a*) in a cross-section of a shell of the chambered nautilus, *Nautilus pompilius*, a marine mollusc, showing the spiralling septa which divide the shell into chambers (when water is pumped out, the animal adjusts its buoyancy with the gas contained in the chambers); (*b*) water whirlpool; (*c*) whirlpool galaxy; (*d*) pine cone with spiralling bracts

metamorphose various shapes into one another (Thompson 1942).

Alexander Lauterwasser exposed water to vibrations of different frequencies. He demonstrated that at certain frequencies the patterns of standing waves arising on the water surface produced striking images of beauty resembling various forms in nature, including a multitude of plant forms, snails, sea urchins and jellyfish (Lauterwasser 2003).

The hidden nature of rhythm and movement within water

The Austrian forester and researcher Viktor Schauberger studied the influence of water vortices on buoyancy and on the quantity of outflow of rivers and channels. He has shown that the swirling eddies in mountain streams enliven the water and give it additional properties not yet fully understood. He managed to improve timber flotation methods with special shapes of the flumes (artificial water channels) and appropriate temperature controls. Schauberger also studied the influence of minute changes of temperature in water, which were ignored by hydrology and hydraulics (Schauberger 1998b, Alexandersson 1982, Bartholomew 2003). He showed that living water is composed of many layers with subtle variations in electrical charge, temperature and density, which influence its pattern of movement and physical properties (Bartholomew 2003).

Therefore when water, comprising such strata of different temperature and densities, flows down a river bed, these layers travel alongside and above each other for a long period without mixing. Schauberger looked for a deeper understanding of principles of fluid motion; he wondered why the fluids in a chicken's egg circulate without a heart, and he observed spontaneous whorls and upsurges of water in lakes (Schauberger 1998a). Schauberger also worked on different forms of energy related to oscillating water and air, and the principle of 'implosion' (Schauberger 1955, 2000, Alexandersson 1982).

He was deeply convinced that conventional science 'thinks an octave too low', and because of its materialistic approach neglects the underlying energetic basis for all physical manifestation and has lost sight of the integrated whole (Coats 1996).

The results of Schauberger, Schwenk, D'Arcy Thompson and others indicate that within a volume of moving water membranes or veils are generated which are invisible under normal circumstances. These are exceptionally rich and complex, for instance, within fresh, unpolluted spring water. And moving water consists of layers that continually flow past one another at varying speeds (Schwenk 1989).

This capacity for complex movement can be equated with water's ability to support life. Technically processed and polluted water loses this quality of movement and may not exhibit it even after purification treatments, when it is deemed pure and potable (Schwenk 1967).

The layers of veils and membranes generated in moving water could be regarded as a reflection of membrane processes that are fundamental in living cells and organisms. It seems evident that rhythms are intimately related to life processes: all living organisms incorporate and express rhythms in different ways.

Water is the carrier of rhythm and as such the mediator between organism and environment. Using treatments for inducing appropriate rhythmical movement in water, we need to establish how water can again recover its capacity to support and enhance living processes in nature (Wilkes 2003).

Schwenk (1967) also demonstrated that water responds very sensitively to cosmic influences, for example to the arrangements/movements of planets in the heavens. This mediating effect for cosmic influences and formative forces can only be established if water is not constricted in its natural movement and rhythm. The rhythmic patterns of waves and meandering rivers do indeed show that water flows rhythmically by nature.

The word 'rhythm' is derived from the Greek verb 'to flow'. Generally, rhythm should be distinguished from frequency. A rhythmic phenomenon is not characterized by mere repetition of itself (such as a monotonous beat), but is a combination of long- and short-period vibrations, thus leading to increasing and decreasing pulses. The dominant or fundamental frequency is modified by harmonics of higher frequency, resulting in a complex rhythmic pattern. The overtones of a musical note offer an example of such harmonics (Nick Thomas 1983).

Self-similarity and geometrical approaches to pattern formation in nature

In recent years, science has explored various mathematical and geometrical approaches to a deeper understanding of pattern formation, such as the forming of a crystal by freezing water, or the formation of patterns in nature that show principles of self-similarity.

Many forms in nature reflect such principles of self-similarity, in that the overall shape of an organism or part of an organism, such as a snow crystal or a fern leaf, reappears as a similar pattern on a smaller scale within its structure. These self-similar repetitive patterns, which could be considered as a rhythmical process 'frozen' in space, arise in clouds and in water as vortex-like forms (Fig. 1), as well as in ice and rocks (Fig. 2, a and b). In the animal kingdom, such patterns are pronounced in shell cross-sections (Figs 2c, 3a, 7).

Examples of highly self-similar forms in plants include the broccoli variety 'Romanesco' (a cross between cauliflower and broccoli), cow-parsley flowers and fern leaves (Figs 4 and 6). In the cow-parsley flower, self-similar shapes appear at three different levels. And in the form of the 'Romanesco' plant, each floret is a similar but smaller version of the whole plant; and self-similar forms can be detected at at least four different levels within the overall form (Fig. 4). The structure of growing plants often resemble patterns of water and air in motion, and many plants resemble upright 'paths of vortices' (Schneider 1995).

In searching for an understanding of complex forms and processes in nature science has increasingly explored fields such as dynamic systems, fractal geometry and chaos theory, as well as principles of self-organization and complex generative patterns (for example Bohm 1980, Briggs and Peat 1984, Peitgen et al. 1992a, 1992b,

Fig. 4 *Cow-parsley flower* **(above, left and right) and** *broccoli 'Romanesco'* **(opposite page) showing the principle of self-similarity. The overall shape is repeated at least three times at smaller scales in the cow-parsley flower, and at least four times in the 'Romanesco' plant, with a spiral arrangement of florets.**

Peitgen and Richter 1986, Peitgen and Saupe 1988, Sheldrake 1988).

According to modern chaos theory, life occurs at the boundaries between chaos and order, whereby so-called 'strange attractors' are formed, which indicate turbulence. Turbulences and vortical movements appear, accordingly, to be enhancing and life-sustaining qualities (Gross 2000b).

Investigations into the field of projective geometry and chaos theory have opened pathways towards describing, geometrically and mathematically, the complexity of organic and inorganic forms (for example Edwards 1985, 1993, Peitgen et al. 1992a, 1992b).

Mathematical structures such as the Mandelbrot and Julia sets bear a strong resemblance to organic forms and demonstrate self-similar patterns at many levels up to infinity (Peitgen et al. 1992b, Peitgen and Richter 1986, Peitgen and Saupe 1988).

A branch of fractal and chaos theory is concerned with how complex shapes, such as snow crystals, fern leaves or tree shapes can be generated geometrically and mathematically.

One system that was developed reproduces transformed and reduced images until a new pattern of growing complexity emerges (Peitgen et al. 1992a, 1992b). Several specific transformations (using scaling, rotation and translation) repeatedly applied to the overall shape by an iterative process can result in complex patterns bearing a strong resemblance to snowflakes, leaves or trees (Figs 5, 6).

The starting point is any image or geo-

metrical shape; the resulting image is determined only by the number and kinds of transformation (e.g. a triangle in Fig. 5a or a trapezoid in Fig. 5e at the first level).

For example, a two-dimensional image of a snowflake can be obtained by repeatedly applying four specific transformations of the original image using scaling, rotation and translation processes (level 2 in Fig. 5 a, e and Fig. 6A a, e). Then, at the next level (level 3, Fig. 5 b, f), each of the four transformations is replaced by the corresponding transformation of the previous total image, thus producing several reduced and transformed copies of the original image in different positions. At level 3, the overall shape consists of 16 forms already, at level 4, 64 forms, and so

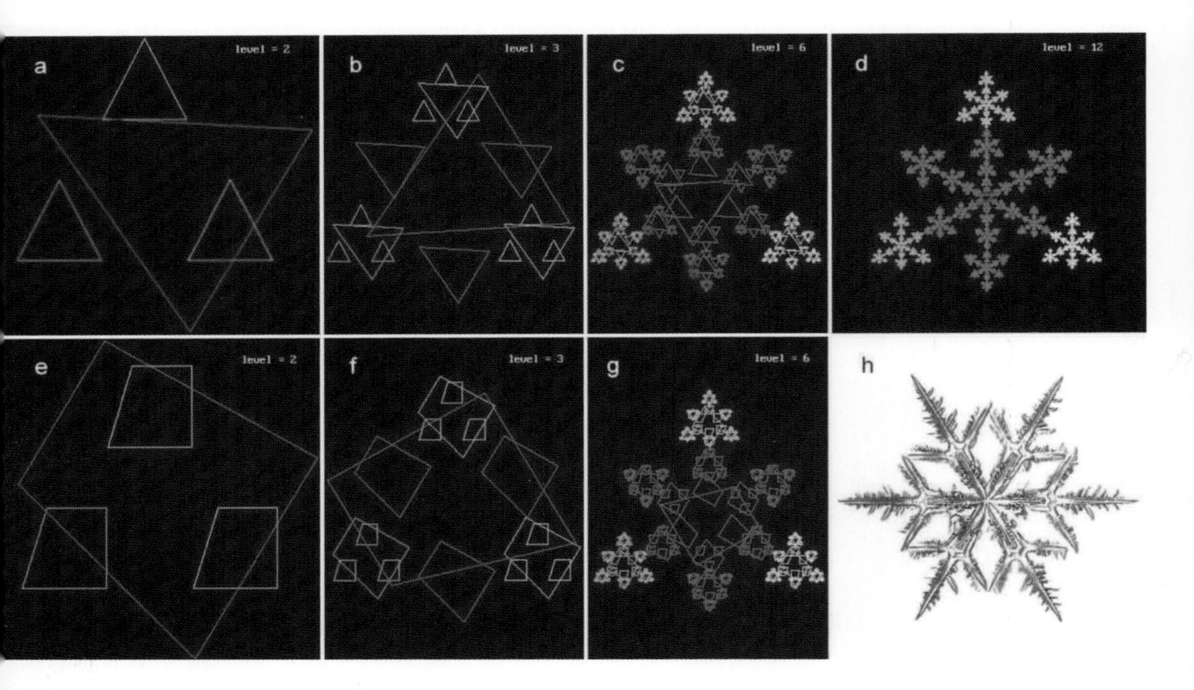

Fig. 5 *Repetitive geometrical process by which complex shapes such as snowflakes can be constructed.* The starting point is a set of transformations of a simple initial form, such as a triangle. If the transformations are the same, then any initial form like a triangle, trapezoid or flower results in the same final image!

This is shown in *e*, *f* and *g*, where the same set of geometrical transformations is applied as in *a*, *b*, *c*, *d*. Here however the starting form is a trapezoid. At level 6 (third from left) the shape is already converging towards the snowflake, but the initial forms (triangle or trapezoid) are still visible in the innermost area of the snowflake form (compare *c* to *g*). At level 12 (see *d*) the two sets have become indistinguishable, triangles and trapezoids have become invisibly small, and the final form has emerged. (*h*) Picture of a real snowflake taken with the light microscope.

on. This process is referred to as a 'Multiple Reduction Copy System', since it produces sequentially reduced copies of the original transformation, leading to images with increasing complexity.

A perfect two-dimensional image of a fern leaf can be obtained by repeatedly applying four specific transformations (Fig. 6A a–d), and five specific transformations produce a tree shape (Fig. 6A e–g).

Interestingly the resulting form, as already stated, does not depend on the initial shape (it may be a triangle, rectangle, a flower shape or any other form), but only on the number and kind of transformations applied. The more often the whole transformation process is applied, the more detailed and structured the final form becomes (Figs 5 and 6A).

The share of single transformations

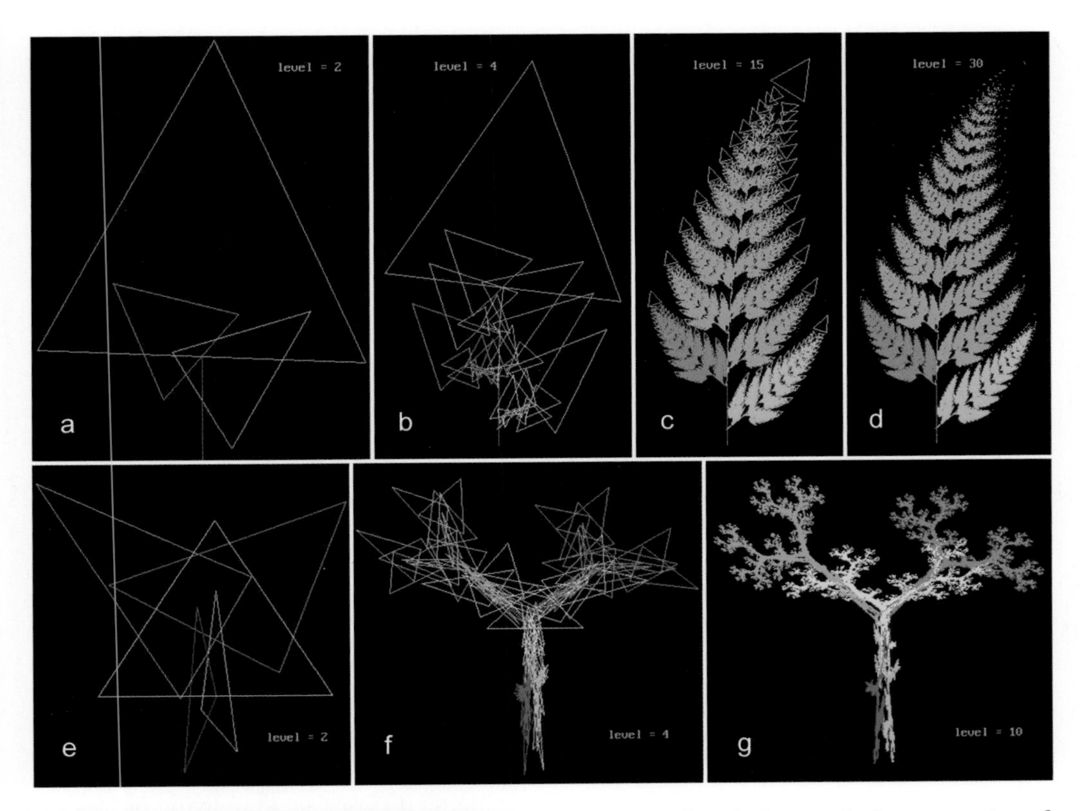

Fig 6A (above) *Geometrical construction of two-dimensional shape of a fern leaf (a–d), and a tree (e–g)*, by repeatedly applying scaled and rotated transformations of an original form, producing reduced transformed copies of the whole. Four transformations shown in *a* applied to the overall image in a continuous process result in a complex fern-leaf shape *d*, and the five transformations shown in *e* lead to a tree shape *g*.

Fig. 6B (left) Fern leaf

within the whole is shown in different colours. Each individual transformation contributes to an entity complete in itself within the whole, but it needs all the other transformations as well to be able to create its own individual structure within the whole.

Complex forms can therefore arise through repetition of a very specific, structured process. Cell divisions leading to the growth and differentiation of an organism could also be seen as the repetition of such an ordered process rather than just as the addition of single cells.

This geometrical approach, where structures and patterns are more significant than individual parts and building blocks, is in harmony with the new scientific paradigm that originated in modern physics.

Such a holistic or ecological approach starts with the realization that the universe need not be seen mechanically as composed of fundamental building blocks, but rather consists of a network of relations. Parts can no longer be separated but need to be understood as patterns in an interconnected, dynamic whole (Capra 1991).

Projective geometry and path-curve surfaces

George Adams emphasized that all physical effects can be described as 'centric forces', since they ray out from centres: centres of gravity, electric charges, magnetic poles, sources of radiation, etc.

In contrast, he also characterized forces working inwards from the whole circumference, from an infinite periphery (Adams 1965). Adams described these forces with the help of projective geometry, path curves and the concept of counter-space. The Austrian philosopher and scientist Rudolf Steiner pointed out that many living phenomena can be related to these peripheral forces raying inwards (Steiner 1975).

In projective geometry, the polarity of points and planes gives rise to an underlying process of 'contraction and expansion' – a principle we find in many organs and vital functions. Adams discovered the path-curve geometry described by Felix Klein and Sophus Lee in the nineteenth century. Its very nature led him to believe that it was linked very closely to life forms. He found for instance that eggs were normally path-curve surfaces and thus began to describe this work to colleagues. He moved on to thoughts about linking water movement to this geometry, and whether it would be possible to improve its quality by bringing it closer again to living processes.

Lawrence Edwards (1982, 1993) later showed that the shapes of plant buds and cones could be characterized geometrically with considerable precision. He discovered that many forms in nature were path-curves, such as pine cones, flower and leaf buds, and at certain stages the heart forms and the uterus. He also showed that forms such as the vortex, and egglike or cone shapes, are mathematically closely related, and can be transformed into each other by changing only one parameter, Lambda (L in Fig. 8).

Prigogine calls forms like the vortex, which are far from equilibrium, 'dissipative structures' (Prigogine 1980). To maintain their shape, these structures must have a permanent input of energy and new

Fig. 7 *Spiral forms:* left, pine cone with vortically spiralling bracts; centre, shell sections with vortex shapes in centre and on outer surface; right, water vortices (vortex below created by higher rotational speed of water)

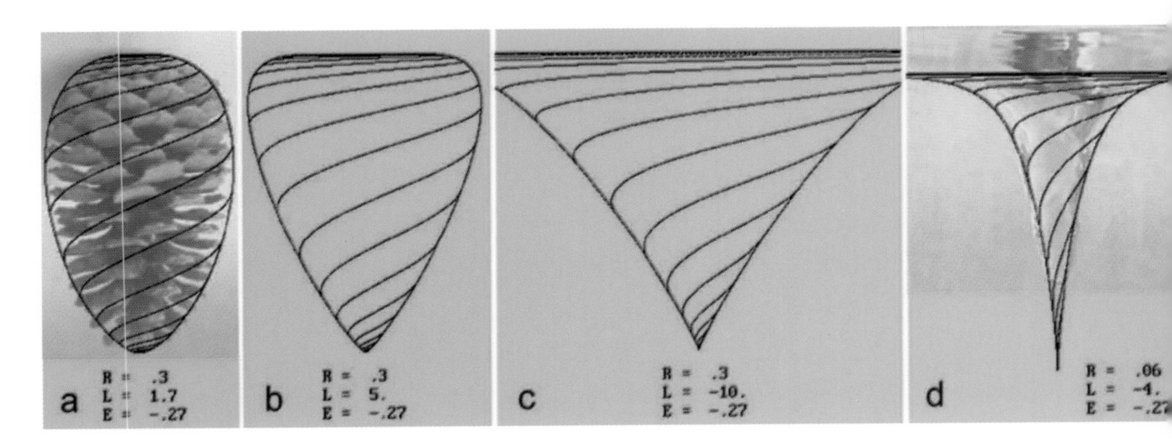

Fig. 8 *Transformation from egglike into vortical shapes* from left to right below. The pine cone shape (see Fig. 7) is modelled with path-curves and can be transformed into a vortex by changing only one parameter, Lambda (L in *a–c*). The water vortex shape in Fig. 7 (right) is obtained by changing a second parameter R, which is responsible for the width of the form (*d*)

material, and at their edge there is a constant exchange of one flow into another. The vortex can survive only by remaining open to an exchange of matter and energy with the environment, which means that the vortex remains stable only by continually flowing (Briggs and Peat 1984).

A whole new field of work and research has grown out of the phenomenological approach to nature (Goethean science), in which the scientist does not separate himself from the object observed, but through participation allows deeper levels of understanding to arise. The basis of this can be found in Goethe's scientific work (Goethe 1993, republished).

Rudolf Steiner and other scientists developed Goethe's ideas further and applied them to find deeper understanding of nature's processes (Bortoft 1996, Edwards 1993, Seamon and Zajonc 1998, Steiner 2000).

The scientific community is giving greater attention to phenomena of rhythm, such as contraction and expansion and their effects, as it becomes increasingly evident that rhythms play a major role in all human activities and in all life processes.

The central question of the Healing Water Institute is how mankind can effectively reintroduce nature's rhythmic, enlivening processes into captured water and water weakened in nature, in order to enhance its life-supporting qualities.

Moving water is intimately connected to both inner or outer surfaces. Water influences and shapes the surfaces over which it flows, and is also influenced by them. Multiple inner surfaces are also created within its own volume. Rhythm and surface are therefore intimately connected and need to be investigated in relation to one another. Such questions led to John Wilkes' discovery of the Flowform principle, which induces a rhythmical figure-of-eight or lemniscate movement in water.

2. FLOWFORM WATER TREATMENT

Human subtlety will never devise an invention more beautiful, simpler or more direct than Nature can, because in her inventions nothing is lacking, and nothing is superfluous.

Leonardo da Vinci (1452–1519)

Development of the Flowform principle

The Flowform principle of generating rhythms in streaming water was first discovered in 1970 by A. John Wilkes, working independently at the Institute for Flow Sciences in Herrischried, Germany.

In its natural environment, water expresses a multitude of different rhythms, such as the rising and falling of the tides, and the rhythmic patterns of waves rolling to the seashore. By its inner rhythmic nature water also shapes and imprints forms into its natural surroundings, through the meanders in a river or a river delta, or rhythmic structures in sand. Cavities similar to Flowform vessels can occasionally appear through natural erosion processes, as at Ayers Rock in Australia (Fig. 9). There is far less order in the rhythms of outer nature compared with the consistency of rhythms within living organisms.

Regions of laminar, harmonic and turbulent flow apply to all three images in Fig. 10. Rhythmic movements characteristic of life processes take place in the harmonic region, which are fundamental to the Flowform method.

Water, as the basis of living fluids within plants, animals and human beings, also flows rhythmically with or without the regulation of the heart. Wherever life and living forms are, rhythm also arises.

The need for a better understanding of water's rhythmic, energetic and life-supporting qualities requires the conscious development of an imagination that combines scientific rigour with artistic sensibility and sensitivity to nature's flowing, living processes.

Flowform designs were developed by Wilkes through his artistic, intuitive study of rhythmic and metamorphic processes in nature, with the aim of aiding water's life-supporting capacity (Wilkes 2003).

All Flowform designs, arising from this long study of nature's creative secrets, combine the highly efficient rejuvenating activity of a mountain stream with life rhythms essential to and inherent in nature.

The unique signature of Flowform water movement is a pulsing figure-of-eight flow pattern, typical of liquid flow within many living organisms. It is worth noting however that this flow pattern occurs only very rarely in an ordered manner in outer nature, and is really only visible to the human eye in Flowform vessels, where the proportions enable consistent rhythms to appear.

Experimenting with the proportions of

Fig. 9 ***Streams of water after two days of rainfall at Ayers Rock in Australia.*** **The colour of the stone turns purple after rain. Water flows into huge cavities similar to Flowform vessels, but created by natural erosion. Some are over 25 metres wide.**

vessels aligned down a slope, Wilkes discovered that rhythm is generated by degrees of resistance and proportion. Surfaces and openings cause resistance to water when it is gaining momentum flowing down a slope. The phenomenon is similar to that of waves forming as water spreads out over a sloping road.

Studies of experimental paths of vortices led Wilkes to consider the process of internal stream impulses in relation to symmetry, asymmetry and proportion (see Figs 11 and 12 below).

All independent living organisms with very few exceptions have an axis of symmetry with respect to their outer physical form. At the same time they all have streaming processes that tend to lead to asymmetrical forms. The heart, which is most related to streaming processes in the

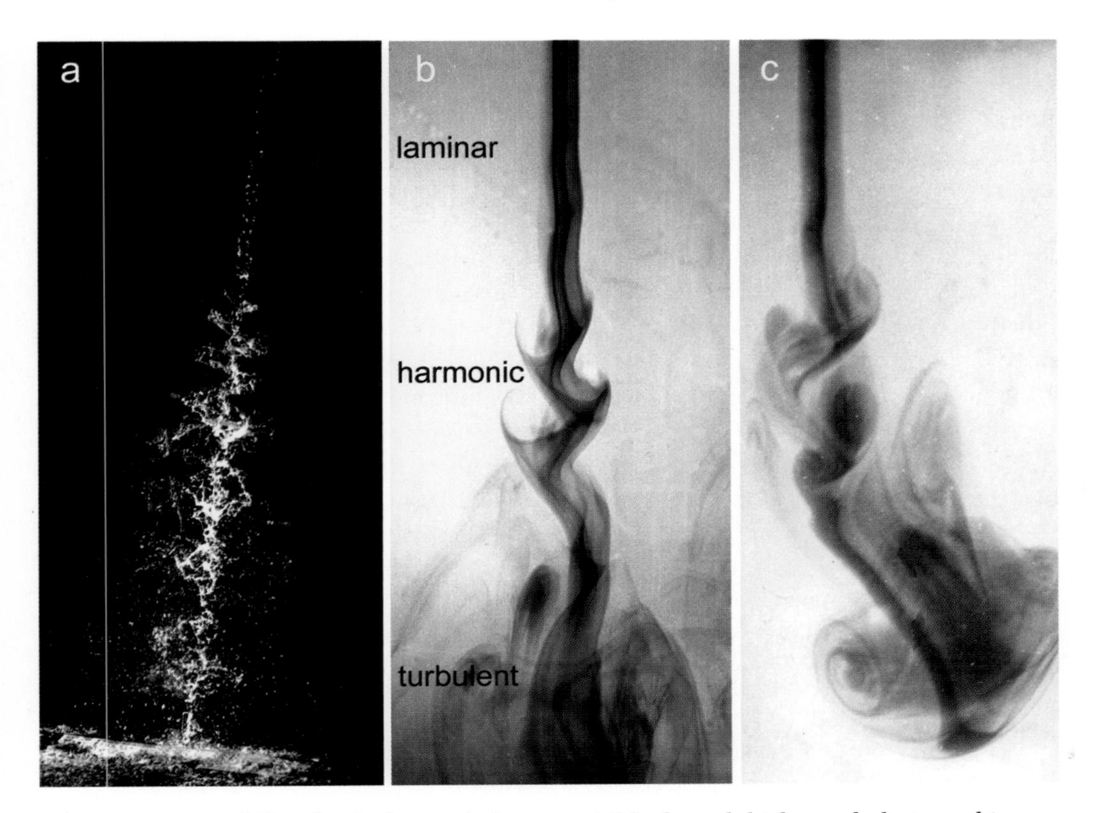

Fig. 10 (*a*) Waterfall – the rhythms only become visible through high-speed photographic exposures (University of Trondheim). (*b* and *c*) Paths of vortices with a rhythmic pattern created by a horizontal flow of coloured water into still water through an opening, (*c*) A structure resembling limb bones with a joint.

Even in regions of turbulence and chaos, structures of a higher order can arise, such as the vortex, lower right, with the bonelike structure.

body is asymmetrical both in shape as well as in its position within the body. Thrust processes in fluids lead to symmetry (as for example the thrust of a wave passing through a dyke, Fig. 16 left), whereas a continuous streaming process leads to asymmetrical shapes (as demonstrated with coloured water streaming continuously into still water, Fig. 16 right).

Wilkes thought that symmetry in life forms could be considered as a higher level of being and he posed the question as to whether the use of symmetry in connection with streaming water could lead to a higher condition in its quality (Wilkes 2003).

Wilkes made a symmetrical channel of varying proportions with a smooth slope, and introduced flowing water through it. At a specific place an unexpected rhythmical oscillation occurred.

Thus the Flowform principle was discovered in April 1970. By verifying that such oscillations could also occur in asymmetrical forms, it was shown that proportions and not symmetry lead to rhythms and eventually to larger-scale lemniscatory movements.

It became immediately evident to Wilkes that the path-curve surfaces George Adams had been creating to influence water quality could be introduced into this newly discovered method of generating rhythms in streaming water. Lawrence Edwards confirmed that path-curve surfaces are intimately linked to vortical shapes, which are incorporated into Flowform designs (Edwards 1993, see Fig. 8).

Hitherto, however, no way had been found to enable water to spread out sensitively over such surfaces. Now, with the Flowform principle, rhythmic, lemniscate movements could be used to enable the water to absorb information relating to these special, organ-related mathematical surfaces (Wilkes 2003).

Thus Flowform vessels provide not only a means of artistically demonstrating the phenomenon of rhythmically pulsing water but also facilitate a wide range of applications that – through the rhythmical movement of water – influence biological and botanical processes, since all life processes are themselves always rhythmic. This represents a harmonious combination of artistic and technological applications.

Water streams repeatedly through each Flowform vessel, oscillating from side to side in a figure-of-eight path and thereby multiplying the actual distance the water travels by as much as 10 times that of the length of the physical Flowform cascade, while also receiving rhythmic impulses that are the signature of life itself. In this way the mountain stream and the living pulse, as nature's own methods for improving water's capacity to support life, are combined in a technology derived from a close study of nature's creative processes.

If this rhythmical process can help to imbue water with the necessary information to support life-sustaining qualities, then healing, harmonizing and enlivening processes might be dramatically enhanced (Wilkes 1993).

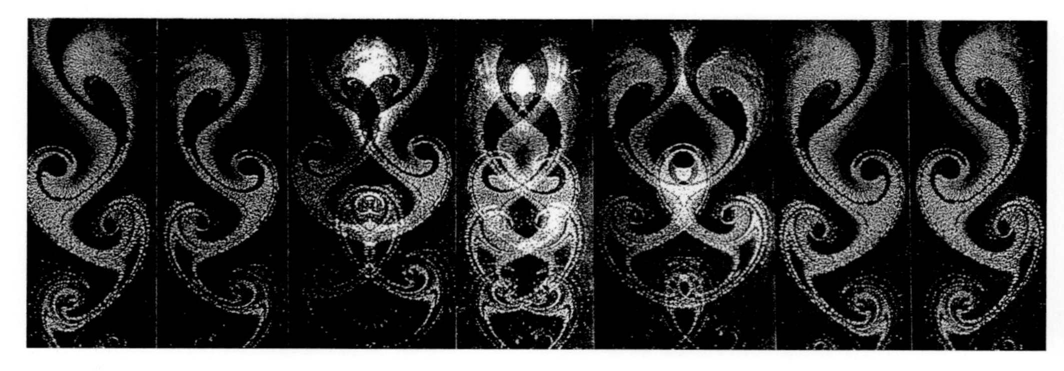

Fig. 11 ***Path of vortices:*** **Photographic superimposition of a mirrored picture moving from left to right. As the mirrored images (first two images from the left) are moved over each other in images 3, 4 and 5 (from left to right), the two paths of vortices overlap, creating a striking similarity to organ-like forms such as the vertebral column**

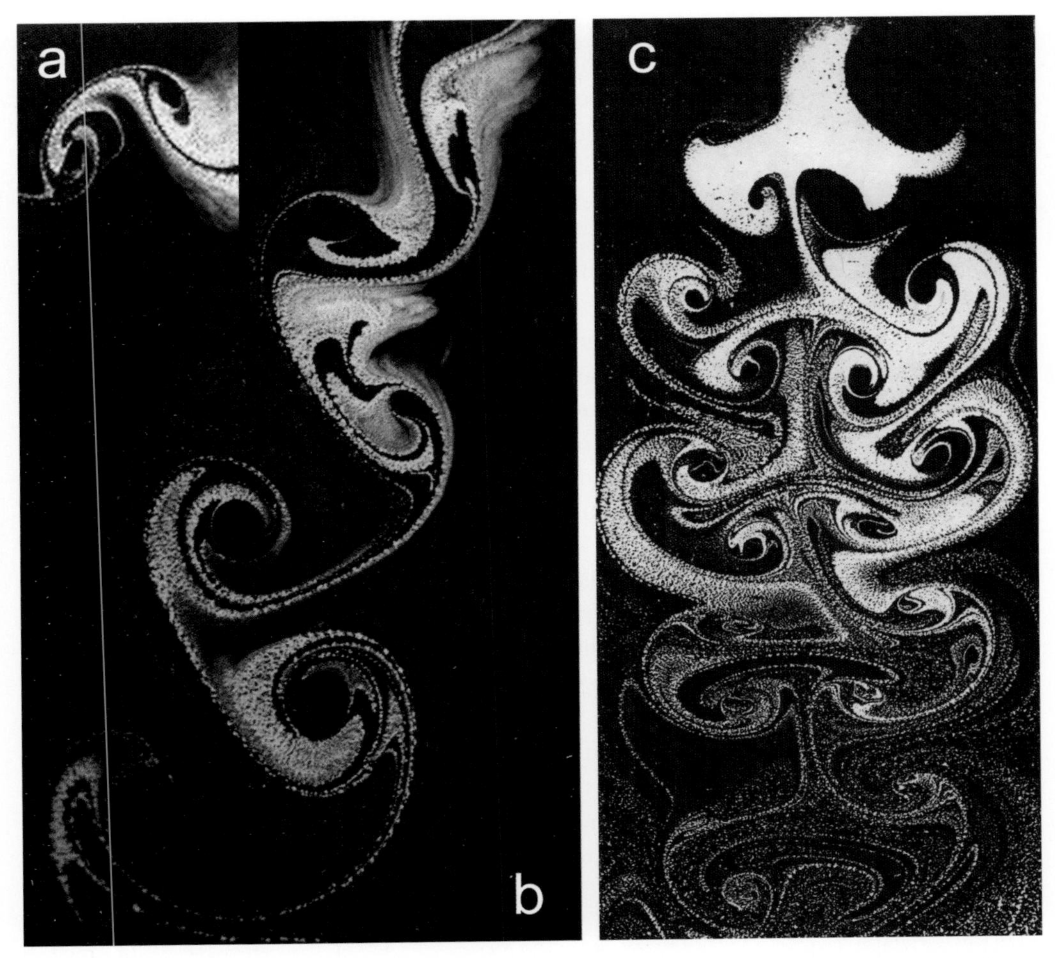

Fig. 12 *Experimental demonstration of paths of vortices*, The asymmetrical pictures can show embryo-like (*a*) and plantlike forms; (*b*) asymmetrical and (*c*) symmetrical paths

Figs 11 and 12 provide insights into formation of pattern in nature. When an object such as a thin paintbrush is pulled horizontally in a straight line through shallow water containing about 60% glycerine to slow down the process, an intricate path of vortices is revealed on its surface (the surface is dusted with a fine powder of *Lycopodium* spores).

The asymmetrical path of vortices on the left in Fig. 12 was created with one brush, while the symmetrical form on the right arose from two brushes pulled parallel through the water in a straight line but with a short delay between them (Wilkes 2001). Speed, viscosity and depth had to be very carefully adjusted to obtain a significant form (Schwenk 1996).

Flowform design research

Since these first developments in the early 1970s, a multitude of Flowform designs have been created. These diverse Flowform shapes all share the figure-of-eight (lemniscate) streaming water flow with a pulsing outflow, unique to the already described Flowform principle. There are single, two, three and four cavity designs, all generating increasingly complex and interesting rhythmic movement patterns.

Most common are the **double-paired** Flowform vessels, consisting of two cavities intersecting side-by-side, either symmetrically or asymmetrically.

The **single-cavity** Flowform design is a later development from original research notes, and essentially consists of one side of the paired cavity form, thus creating a simpler rhythm.

Whereas it is interesting to compare the double-paired Flowform vessel with the human larynx (Schwenk 1996), the single cavity Flowform vessel may be compared with a fish-heart section (Kilner 1984).

Dr Philip Kilner suggests that the Flowform design could be an appropriate model to help reveal certain aspects of heart function. He mentions that in the fish, as well as in the human embryo, rhythmic pulsation of the heart begins before any chambers develop, with the pulsing happening first in the tiny bend and swelling of the heart tube (Kilner 1984). Again this is an area that concerns proportions.

Such double-paired Flowform vessels can be designed and constructed either as single repeatable units to form cascades (Figs 13 and 15) or as metamorphic sequences in which forms change their shape within the total cascade process (***Sevenfold Cascade*** in Fig. 14 left).

The ***Emerson*** model (Fig. 13) is a larger open vessel and was designed originally so that water could spill over the front edge in a waterfall. It is intended for mostly aesthetic purposes. The ***Malmö*** model (Fig. 39b) is a form designed for larger flow

Fig. 13 Symmetrical *Emerson* (left) and asymmetrical *Olympia* (right)

Fig. 14 ***Sevenfold Cascade (left) and Ashdown (right).*** **The size differentiation in the Sevenfold Cascade creates differing rhythmical frequencies with varying gradients and the same flow rate, thus creating a richer movement experience.**

rates, which can be used in sewage plants. The ***Akalla*** model (Figs 15 and 17B, left) has a robust, boulder-like shape and comes in three sizes to express calmer movements in the larger forms, and more vigorous movements in the smaller forms, both using the same flow rate while at the same time accommodating varying gradients.

The shallower forms were mostly designed for aesthetic purposes: these are the ***Emerson, Amsterdam and Ashdown***.

The **radial** design is a later development: here water rises in the centre and flows out in three directions, each at 120° angles to each other, with the three Flowform vessels incorporated into the overall circular design.

One example of such radial design is the ***Ashdown*** form, which is smaller and formed in a contracting and expanding circular pattern (Fig. 14 right). The ***Amsterdam*** and ***Viktors*** models are large radial forms as much as 3 metres in diameter, and are intended especially for parks or townscapes.

John Wilkes found that Flowform designs with identical flow rate but varying size generate a wider range of rhythms within the cascade. Larger models generate slower, calmer and more two-dimensional movements, while smaller shapes generate faster, more vigorous and three-dimensional action (Wilkes 2003). The *Sevenfold Cascade* is an example of a metamorphic sequence of shapes within a cascade (Fig. 14).

The ***Glonn*** (Fig. 15) and ***Vortex*** designs (Fig. 22, left) are two of the double-paired type, which can be built vertically for smaller flow rates (up to 80 l/min) or in-line for larger flow rates (up to 250 l/min) (Wilkes 2001). The deeper *Glonn* forms produce a vigorous movement for treating water used in food processing, such as grain-milk production and baking, and for the processing of drinking water. The *Vortex* design with two outlets in the base takes larger flow rates, produces more vigorous movement and is not intended so much for aesthetic purposes but rather for technical

Fig. 15 *Flowform types: Herten* (top left); *Akalla* (top right); *Järna* (bottom left); *Glonn* (bottom right)

applications such as biological sewage treatment, biodynamic mixing, irrigation processes, drinking water for cattle, and food processing systems such as grain washing.

The main intention within the design of Flowform vessels is to give water the more elevated pattern of movement which exists everywhere in the living world, namely a consistency of rhythm. In outer nature water also moves through the influence of gravity but is continuously modified by a multitude of influences, which ultimately lead to a chaos of rhythms. Thus the pulsing lemniscatory condition is sought.

Another design method involves the inclusion of specific mathematical path-curve or organic surfaces, which can be built into the system bringing another quality of influence.

These two design methods could be called

1. **Scientific Imagination:** an empirical approach based on the designer's deep understanding of water movement, metamorphic processes in nature and sculptural design.
2. **Mathematical:** based on path-curve mathematics; this approach transposes specific shapes that optimize life-supporting influences in nature (Adams 1965 and Edwards 1993) onto Flowform sensitive surfaces to potentially influence water as it flows over them.

These aspects need to be applied once the more aesthetic and also functional choices have been made with respect to materials, location and purpose. The design research certainly has a high potential for ascertaining and optimizing specific applications, such as drinking water purification, specific mixing or energizing applications, and artistic installations.

Three functions are carried out by all Flowform designs: they generate rhythms, they mix, and they oxygenate. There may well be additional functions such as precipitating particles or detoxifying.

Over time, Flowform designers have noticed that overall shapes can echo natural biological organs and their functions. The following gives some ideas but conclusions can also vary. This shows how much the invention has arisen from nature's own methods, as a prime example of bio-mimicry achieved long before that concept was formulated:

- **Kidney shapes**: observations and practice show that these mix and stir liquids most effectively, while also retaining oxygenation and rhythmic influences. Examples are the *Järna* (Fig. 15 bottom left) and *Vortex* designs (Fig. 27).
- **Heart shapes**: these clearly develop strong pulsing rhythms in a more focused manner than other designs. We are considering their specific use in energetic information exchange. Examples are the *Malmö* and *Glonn* models (Fig. 15 bottom right).
- **Lung shapes**: these forms are wider and more open, and maximize air-water surface contact. Such 'lung' forms are for example the *Emerson* (Fig. 13 left), or the large *Olympia* (Fig. 13 right) and *Akalla* (Fig. 15).
- **Liver shapes**: these are represented by the single cavity designs, which have nevertheless been seen as relating to more primitive heart forms in fish. The resulting rhythms are of a simpler single pulse compared to the two cavity forms, which usually demonstrate an unequal double pulse.

Naturally, there are also combinations of shapes with transitions that create a mixture of influences. Thus a higher effectiveness can be achieved through the metamorphic series that includes all such forms and their functions in one continuous Flowform cascade (for example *Sevenfold Cascade* in Fig. 14 left).

Materials in Flowform production

Flowform manufacture utilizes a wide variety of materials. All materials will have an effect upon the water itself, from a qualitative and also quantitative point of view. This can be chemical, electrical or even functional with respect to the friction of a surface having an impact on flow rate and rhythmical properties. If we take into account the extreme surface smoothness of materials such as glass, ceramic or plastic, this can make the function extremely sensitive.

All elements and their combinations emit frequencies at a molecular level, and these can be utilized to influence the cluster structure and information signature of water flowing through Flowform materials specifically used for this water treatment purpose. Interesting research into ceramics and energetic stones is emerging in recent years that we intend to incorporate into our choice of production materials.

These material effects can either form part of the treatment or be an aspect we seek to avoid. Materials also affect the nature of the design. Further research on the suitability of materials therefore is of paramount importance.

Concrete, and the sophisticated **reconstituted stone mixes**, have been used extensively. The highest percentage of natural material at the surface (such as sand or stone) would generally be most desirable, thus excess cement is normally etched away with dilute acid. This material has economic and craft-shop benefits but has issues regarding weight, maintenance and efflorescence.

Ceramic has been used for relatively small-scale work. It is possibly the best with respect to neutrality. Stoneware or porcelain is preferred, and the huge variety of glazes and clay bodies provide a great range of artistic choice, with potentially varying frequency influences. This material is also convenient for large-scale mass production where facilities are available. Some of our work is being outsourced in SE Asian ceramic facilities with very high skill levels, drawing on thousands of years of uninterrupted manufacture combined with modern environmental methods.* It has been found that glazes (with varying metallic content) can be used to advantage for specific purposes.

Glass has been used effectively but is expensive to produce in small quantities. The potential for recycling glass is attractive, combined with recent technical innovations with varying glasses for different functions.

Thermoplastic is very useful for vacuum forming and drape techniques, rotational production and for less expensive solutions. There are types of plastic that can be inert with regard to material and vapour loss. Some environmental analysts

* This work is run under ethical and properly waged conditions, with an education fund for workers' children and numerous forms of community support.

see 'plastic' as a positive long-term material because it can be recycled repeatedly.

Thermoset (polyester, vinylester, epoxy resins). Glass reinforced resins can be used with heat treatment to drive off excess undesirable substances. Hollow products can be cast quickly using a proprietary system developed by colleagues in New Zealand.

Composite stone. These same colleagues (Design for Life Ltd) have developed a new casting material in hollow, lightweight, very long-lasting, strong and beautiful stone, made from 75% natural materials bound together with polymers. Research is ongoing into sustainable polymers. This has given rise to a mass production method that is many times faster than concrete casting, easier to freight and install, with far more varied and natural surfaces.

Metals are very desirable from a design point of view and can be used to achieve chemical, electrical and biological effects, but are yet to be thoroughly investigated.

Stone is perhaps the ultimate solution but does present technical, transport and economic problems, which can certainly be overcome if resources are available.

Fig. 16 *Symmetrical pattern*, arising in water due to the thrust of a wave entering a passage through a dyke, bearing a striking resemblance to Flowform shapes (left). Coloured water streaming into still water producing an asymmetrical pattern (vortical meander, right)

Natural polymers are the latest materials to be investigated, especially with respect to their possible positive effect in relationship to drinking water treatment. We are following closely the extensive environmental-impact research occurring worldwide in this industry.

A great deal of useful and interesting research is still waiting to be carried out, despite the fact that design research in conjunction with materials is something we have focused strongly on. For instance, the addition of trace elements to surfaces that can leach into water is one project to investigate. This can influence water frequency (energetics) or also put natural nutrient elements back into water that may be lacking due to human industrial usage or filtration. Likewise the addition of natural frequency emitting materials, such as silver, may influence the water positively.

All materials have an energetic and chemical impact on water quality. Also all materials whether viewed as environmentally acceptable or otherwise need careful analysis, as our emotional preferences may not be backed up by the science of environmental disadvantages or benefits. Sustainability issues are complex. For instance, many people prefer concretes to polymers, yet cement production is a very industrial process which pollutes earth and ground water with heavy metals.

Flowform applications

Since 1970, the colleagues and associates of the Healing Water Institute,* located at Emerson College in Forest Row, Sussex, have installed approximately 2500 installations in some 50 countries.

Flowform cascades are currently used in a wide variety of applications, including: aesthetic installations for public, educational and private purposes; functional installations connected with biological purification, farming, food processing, drinking water, pool treatment, interior air conditioning; plus therapeutic, medicinal, and other functions.

Industry sectors with existing installations are:

1. Interior design
2. Land and water areas
3. Human health and development
4. Water treatment
5. Major public designs

It is beyond the scope of this document to detail all applications. This will be addressed in a further 'project portfolio' collation. Nevertheless, below are some functions of note.

Biological sewage systems

Human community 'black-water' treatment

The combination of lagoons and reed-beds with Flowform cascades is an efficient way

* Whereas the common name is 'Healing Water Institute', at the time of writing the UK legal name is 'Healing Water Foundation'. Official application for use of the word 'Institute' in the UK is underway. The name is officially acknowledged in New Zealand and the USA.

of establishing a biological wastewater treatment system. Bacteria treat waste passing through a constructed reed bed. Aerobic bacteria are associated with the rhizosphere (root zone) of the reeds, and anaerobic bacteria with the surrounding sediments. Within this bacterial matrix, organic wastes, nutrients and a variety of chemical compounds can be broken down and stabilized (Worrall 1992).

The first major Flowform effluent project was commissioned in 1973 for Järna Rudolf Steiner Seminariet, Sweden, and was observed augmenting natural processes over many years. This led to a research project in the 1980s at the Warmonderhof Biodynamic Training Institute in Holland. Since the 1970s many such systems have been built and are functioning successfully round the world.

'Grey-water' systems

Apart from black-water systems such as at Järna and Warmonderhof, Flowform cascades can be used to improve water quality in combination with filters in grey-water systems (non-sewage wastewater from laundry and showers). One example of such systems receiving communal laundry water has been set up in a small community near 'The Channon' in New South Wales, Australia, using a gravel filter and a reed bed in connection with a Flowform cascade. Resulting water can be used effectively for irrigation or returned to open nature in a condition that adds quality to the environment.

Dairy shed effluent treatment

Dairy farming worldwide presents the challenge of environmentally acceptable treatment of wastes coming from milk-producing farms.

Nutrient-rich farm-dairy effluent (FDE), which consists of cow excreta diluted with wash-down water, is a by-product of dairy cows spending time in yards, feed-pads, and the farm dairy. Traditionally, FDE has been treated in standard two-pond systems and then discharged into a receiving freshwater stream or onto fields as irrigation. Using the normal method of sluicing the cowshed waste out into standing ponds that are treated through sedimentation, sunlight and slow pond current movement, the effluent stays anaerobic in the ponds. When the effluent is sprayed out on the farm paddocks, grass grows rank and dark green, collapsing with strong growth. The herd cannot be returned to such fields for a month because of health issues such as mastitis associated with the anaerobic sludge.

Research on the effects of land-treating FDE and its impact on water quality has shown that between 2% and 20% of the nitrogen (N) and phosphorus (P) applied in FDE is leached through the soil profile. In all studies, the measured concentration of N and P in drainage water was higher than the ecological limits considered likely to stimulate unwanted aquatic weed growth.

In New Zealand and Australia, some ten dairy farms have had Flowform technology installed adjacent to their milking shed, transforming its twice-daily effluent wash-out through a Flowform cascade, sometimes in conjunction with biodynamic compost preparation.

With effluent treated in large, open, buried tanks over a two or three-week period, the waste can be transformed successfully into a liquid fertilizer which is then sprayed onto paddocks. The grass is able to take this up as food without stress and cows are able to return to the field within a week.

The practical success of these applications implies that other similar projects would work in other situations, with appropriate fine-tuning.

It is evident that cascades are useful for the introduction of oxygen, apart from the effects of rhythm providing an environment to support micro-organisms, which themselves incorporate rhythmic processes. All living creatures and plants are rhythmic. Fluids move rhythmically within them, even without a heart, as occurs with smaller, simpler animals and of course plants. When water, which is itself rhythmic, surrounds or is drunk by living things it is presumed to enhance their own rhythmic processes. Used in conjunction with suitable water-lifting techniques, damage to micro-organisms is virtually excluded.

Flowform cascades generate a rhythmic environment sensitively related to and supportive of these organisms that are intrinsically rhythmic by nature. These systems indicate ways to support and enhance biological purification processes, revitalize purified effluents and accelerate their re-entry into the natural cycle. They also transform a pollutant by-product into a valuable liquid fertilizer and help the farmer manage his land and avoid regional council fines.

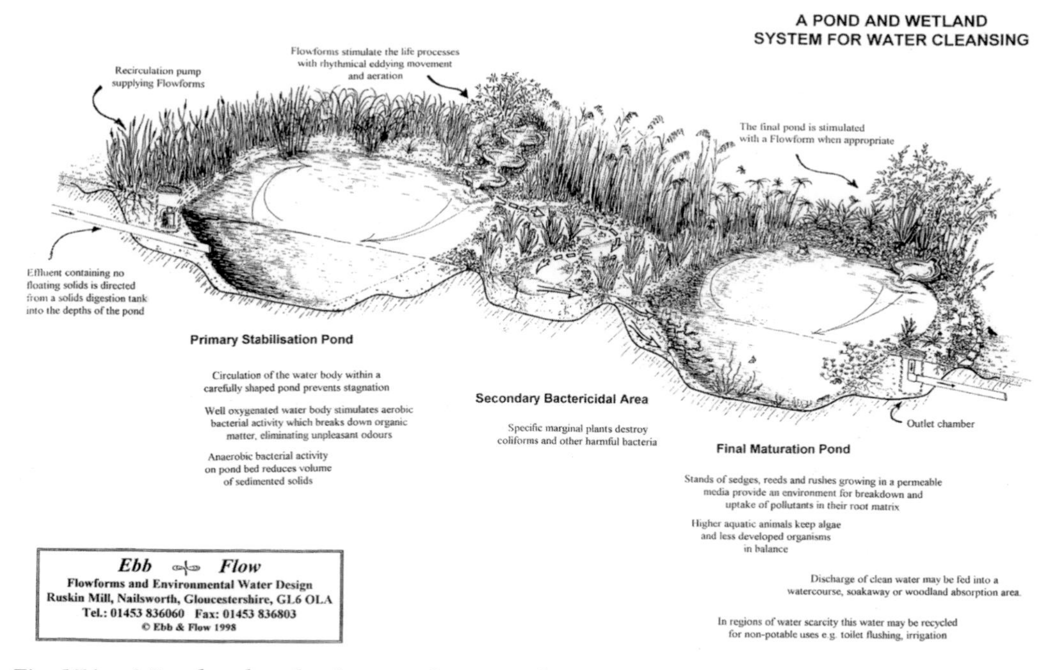

Fig. 17A *A Pond and wetland system* for water cleansing, sketch from Ebb & Flow Ltd, Ruskin Mill

Fig. 17B ***Cascades with lagoons and reed beds*** **in Järna, Sweden (left), and Theedingsweert, Holland (right)**

Other effluent treatment

David Clements worked with Flowform treatment in the 1970s for a pig-slurry project at Broom Farm near Stourbridge, W. Midlands. In this project, the liquid was turned successfully into aerobically transformed 'sweet-smelling' liquid fertilizer.

At a chicken abattoir in Byron Bay, NSW, Australia,* Flowform Vortex design cascades were installed to influence effluent liquids. This abattoir slaughtered 26,000 chickens per week and the washing from the slaughter room went through a clarifier into a 4.3 megalitre pond. From there water was used for irrigation on an onsite turf farm. However the stench caused neighbours to complain and the Environmental Protection Agency were about to close them down when the Flowform project was started. Dissolved oxygen readings were zero to start, which was most unusual. Three sets of seven Vortex models were installed and within two weeks of continuous running the pond's dissolved oxygen was 2.3 ppm. By the end of the sixth week there were readings of 6.3ppm and by then the abattoir's neighbours were ringing to enquire where the smells had gone. The Environmental Protection Agency extended their license, with the added construction of gravel filters and reed beds.

Farming applications

Applications of Flowform cascades for farming are related mostly to mixing and treatments of various biologically catalytic liquids. In Australia and New Zealand especially, Flowform biodynamic stirring is being used extensively on a large scale. There, large areas of inaccessible hilly grasslands are sprayed by air via helicopters and planes.

Many thousands of hectares have been granted Demeter biodynamic certification

*While the ethics of supporting such an operation may be questionable, it does demonstrate the efficacy of Flowform treatment. As the Flowform organization and business progress, ethical policies will be developed relating to the nature of the farms or industrial practices supported by this technology.

with use only of Flowform stirring methods, which essentially adhere to Dr Rudolf Steiner's suggestions for these soil-treatment preparations.

Flowform installations are useful not only for mixing preparations, but are in continual use for treatment of liquid manures, seaweed and fish fertilizer production, irrigation water for plants and drinking water for animals.

Food and drinking water processing

Water quality has become a major concern in connection with food processing. Mostly such water is taken from the municipal supply, filtered and possibly enriched with minerals and then used. However the water has been under pressure moving through pipes, cut off from its natural movement; and while chemically and organically sound it may well not have much capacity to support life energetically.

At Herzberger bakery in Fulda, Germany, water used for baking bread was run over a cascade of granite steps followed by a stack of Flowform vertical Glonn models (Fig. 28). See 'Biodynamic food production and Flowform effects' on page 69 for the results of this treatment.

New Flowform technology was also installed at the Bio-Sophia factory in Lillehammer, Norway, to service various aspects in the production of grain milk: washing,

Fig. 18 ***Flowform water treatment installation*** **for Bio-Sophia in Lillehammer**

swelling, cooking, fermenting and diluting. The 5.25 metre high water treatment unit with ceramic cascades was housed in a protective polyhedron. The form field created within the polyhedron has been shown to have an additional sensitizing effect on the process (Wilkes 2003).

At Giubiasco in Switzerland, drinking water for the municipal supply is gathered from springs high in the mountains. The energy generated by falling through pipes down to the valley is extracted with a turbine and generator before it enters the municipal system. Through this process, however, the water appears to become more aggressive, and was attacking the walls of the concrete reservoir. For this reason Flowform cascades were installed below the turbine to renaturalize the water. The success of this was indicated by crystallization tests (Fig. 19).

Other applications

Landscape/water areas

Clients have reported repeatedly that plants and animals in the vicinity of Flowform cascades respond with greater activity and better health.

Interior design

Flowform vessels can be used effectively for humidifying dry environments, especially during dry northern winters; and the larger cascades with strong waterfalls emit large negative ion impulses.

Health and well-being

There are a variety of uses for medical applications wherever life-supporting liquids are used.

There is a use in homoeopathic production, where hand-mixing of large volumes can be extended/supported by Flowform treatments.

Therapeutic results from patients' proximity to Flowform water movement have been reported worldwide for decades: people with various psychological and physical conditions relax and grow calmer, sleep disorders can be alleviated, and depression and emotional blockages can be relieved. Much more research is needed in this area, particularly as regards physiological responses.

Fig. 19 *Copper chloride crystallization images* from a French laboratory showing spring water from Valle Morobbia (left), after a turbine (centre), and after Flowform treatment (right). The central picture shows high levels of entropy or disorder, while the right picture shows a return to harmonic forms similar to the original spring water. This is a fine example of energetic pollution where the chemical and organic quality of the water remains the same throughout.

Educational

Flowform cascades have been installed in many kindergartens and schools, and observations show that children react most positively to the rhythmic environment. Over-active children tend to calm down and become more focused, introverted quiet children become more engaged, and greater social skills with cooperative language appear to develop.

Flowform playgrounds seem to positively influence diverse sensory development as well, and offers an experience of flowing water to many city children who often have little relationship to nature.

Such play areas are highly regarded as a beneficial educational tool, and also provide a central focus for the school community.

Tap-water treatment

Tap water is often highly mineralized, and after supply to the home or office sink is frequently filtered or distilled. However, filtering or distillation does not change the information frequencies within such water, removed as it is from the health-giving influences of natural movement.

Flowform water features offer a means of returning this life-supporting capacity to tap water through a flow action that nature itself uses to improve water in both mountain streams and within living organisms.

Pond treatment

Numerous observations show that Flowform cascades oxygenate ponds effectively but also change the flora and fauna so that these become similar to those in a stream.

In New Zealand and other countries, Flowform cascades were installed in at least ten swimming pools, some conventional and others specially designed as natural swimming pools incorporating small wetland systems, filtering reed beds and other natural features. This appears to produce a balanced ecosystem, in which the use of chlorine can be avoided completely (Pearsall and Innes 2000).

Generally, these cascades greatly reduced the need for treatment chemicals, while also increasing the water's sparkle, freshness and smoothness. Professional pool managers consider these water quality observations to be as valid as chemical measurements in evaluating actual improvements to water.

Agriculture

Biodynamic preparations: extensive use of Flowform stirring in New Zealand, Australia and elsewhere shows conclusively that this method activates the liquid biodynamic preparations very well. Up to 10,000 hectares were certified as Demeter quality in the 1990s, using only Flowform vessels for stirring (Peter Proctor 1997, Trousdell 1990).

Compost tea stirring: this method of preparing life-supporting liquids for soils and plants is widely associated with Flowform technology.

Plant germination: field and laboratory trials in New Zealand, the UK and Germany have shown that, when started with Flowform water, germinating seeds tend to show overall stronger growth throughout their life.

Morphology plant studies with coriander over three years in NZ in the early 1990s by Menzo de Boom, Hans Mulder and Iain Trousdell showed a more balanced metamorphic leaf sequence in plants germinated and irrigated by Flowform-treated water for the first three weeks of life, in comparison to control plants.

Plant trials with water taken from differing lengths of Flowform cascades in the 1980s in Australia by David Julian and Scott Douglas indicated increasing and decreasing influences from smaller and larger numbers of vessels in a continuous cascade. Numbers over 12 consecutive vessels without an intervening chaos chamber did not necessarily result in continuously increasing life-support effects. This suggests, interestingly, that more research is needed into the effects of chaos rhythms.

Aquaculture trials showed benefits for goldfish and farmed trout in NZ, demonstrating that these fish preferred Flowform-treated water.

Hydroponics: some trials in New Zealand showed that the plants grow stronger roots and foliage in water that flows through a Flowform cascade while traversing growing channels.

Animal drinking water: strong indications with pets and farm animals show that Flowform-treated water is preferred, resulting in increased consumption.

Industrial

Once planners and engineers realize that water has energetic, information-frequency qualities and active freshness, Flowform technology could be used before, during or after processes that reduce water quality in any factory using it for profit.

Special designs could also be used for specific mixing processes of various liquids, with the potential to focus on any or all of laminar, rhythmic or chaotic flows.*

Many other applications have been and are currently being reported to the Healing Water Institute. These accounts need to be fully scrutinized before publicizing them further.

Water-lifting techniques

From the beginning, in 1970, a question of paramount interest was how water could be returned to the top of a cascade without losing any subtle effect that might be gained through the downwards rhythmical treatment. This is a particular concern when the same water is recirculated continually in a cascade.

The negative effects of pump pressures and mechanically induced turbulence can lead to degeneration of water quality. Though in some ways more cumbersome than a conventional centrifugal pump, the Archimedean screw, which lifts water in rotating pockets, delivers water efficiently and is a more benign way of lifting it (Wilkes 2003).

Design research then led to the question whether the screw could be so designed that the process of lifting might contribute something beneficial and, in combination with Flowform cascades, enhance the anticipated effects of rhythmic treatment.

* Laminar flow describes the smooth flow of water, with a laminar set of invisible sheaths overlaying each other, giving rise to smooth movement. The other two types of water movement are harmonic/rhythmic and turbulent/chaotic.

These questions led to the development of a transformed Archimedean screw. This apparatus consists of an open twisting channel wrapped round an axis, to be used for lifting water while carrying it through a series of left- and right-handed vortical movements.

This creates a complex spatial movement that is something like an elongated spiralling lemniscate, and provided a compatible lifting treatment similar to and in harmony with a Flowform cascade.

The **Virbela Screw Pump** is our name for this technology. It offers an ideal opportunity for incorporating path-curve surfaces into water lifting (Wilkes 2003).

Path-curve surfaces of this kind embody, in archetypal mathematical form, the kinds of surface generated by living organisms, to which they are sensitively related. (See p. 30.)

From a geometric perspective we can postulate that if water were to be encouraged to follow the asymptotic curves on such path-curve surfaces, a quality of balance would be achieved in the movement dynamic between space and counterspace. This relates to work initiated by George Adams at Herrischried, Germany (Adams 1965). See also John Wilkes' essay on projective geometry in this volume (pages 97–105).

Summary

Everywhere that water is used by humanity, whether in agriculture, industry or domestically, its quality is inevitably reduced. UNESCO statistics indicate that human beings are presently capturing fresh water from 65% of the world's fresh water sources, and that within 35 years this is expected to

Fig. 20 ***Järna Flowform cascade with Archimedean screw*** **used in the Rural Development programme at Emerson College (left);** ***Virbela Screw*** **development model (right) (Wilkes 2003) here with the help of Don Ratcliff**

rise to 90%, thereby removing vast amounts of nature's water from the potable water cycle, to the detriment both of the water itself and the natural environment.

It is vitally important that decision makers find ways to return this fresh water to nature in a good condition; and where it is not returned to nature it needs to be treated in such a way that its chemical, organic and energetic quality is improved after use.

Flowform eco-technology enables water to be returned to nature, so to speak, while still in a captured condition; this is achieved by running the water through cascades designed entirely in harmony with nature's own optimum methods for oxygenating, restructuring and energetically treating water.

This is the primary research and design purpose of the Healing Water Institute.

3. PATH-CURVE SURFACES AND FLOWFORM DESIGN

It will be needful to find the true idea for any process in Nature, instead of blindly applying ideas that may be foreign to the essence of it.

George Adams (1894–1963)

The possibility of designing mathematical Flowform surfaces to further influence water quality was introduced in the last chapter on Flowform water treatment.

In 1970, when the Flowform method was discovered, it became feasible to combine this with path-curve surfaces. This made possible further research in relation to effects on water, particularly in connection with rhythms. (See p. 30.)

Fig. 21 ***First path-curve Flowform model, 1970.*** **Left- and right-handed sections were specially made based on one of the existing path-curve models. They were put together in relation to one another in order to generate an oscillation in water passing through from right to left. The oscillation was converted into a rhythmic, lemniscate movement which encouraged the water to spread out repeatedly over the mathematical surfaces. Previously, prevailing gravitational and rotational forces had dominated. (Wilkes 1970.)**

However, this Flowform path-curve design research was not taken further until the1990s. In 2009 a working model called the Vortex Cowhorn was mathematically generated, out of a path-curve Flowform prototype made during 1994.

Two types of Flowform Vortex path-curve models now exist. The first one (Fig. 22, left), designed for a maximum of three hundred litres/min, was created empirically by Nick Weidmann in the 1980s. Astonishingly, through his profound experience of water movement and the creation of rhythms within a correctly proportioned Flowform design, he was able to find the quality of path-curve surfaces. We were able to confirm this by means of a computer program specially designed by Nick Thomas. These surfaces are concave.

The second Vortex path-curve design (Fig. 22, right), called the Vortex Cowhorn model, uses mathematically generated path-curve surfaces which incorporate the values of the natural water vortex and a typical cowhorn example interwoven within one surface, calculated by Nick Thomas. These surfaces are convex as well as concave (a

Fig. 22 Flowform Vortex design (left) and Flowform Vortex Cowhorn design (right)

double curved surface). Such surfaces are found everywhere in nature.

The idea behind this is to use surfaces that have an intimate relationship to the processes and substances used in mixing biodynamic preparations, which might lead to an enhanced effect on the soil.

Figure 23 comprises a selection of path-curve surfaces constructed during the 1960s at the Institut für Strömungsforschung. This material was prepared by John Wilkes for George Adams' research. It now provides the basis for aspects of our present investigations.

Construction of the Flowform Vortex Cowhorn design

by John Wilkes and Nick Weidmann

During the 1990s, path-curve Flowform design was begun. After constructing the surface by rotation (see Fig. 24) from one of the path-curve surfaces made for George Adams (see Fig. 23) four left- and right-handed sections were cast, for incorporating into one form (Fig. 25). These were united to create one Flowform vessel (Fig. 27). After the form was tested with water numerous times (Fig. 26) a prototype was finalized (Fig. 27).

In order to create a cascade with such Flowform vessels, we decided in 2009 to transform this prototype into the Vortex Cowhorn model by creating two holes in the base. Now the water could swing out over the proffered mathematical surfaces in a thin film, integrating this particular quality of spatial information (Fig. 28). To indicate the effect of such a design, repeated movements are continued for one hour and various tests carried out, for instance on plant growth (see below).

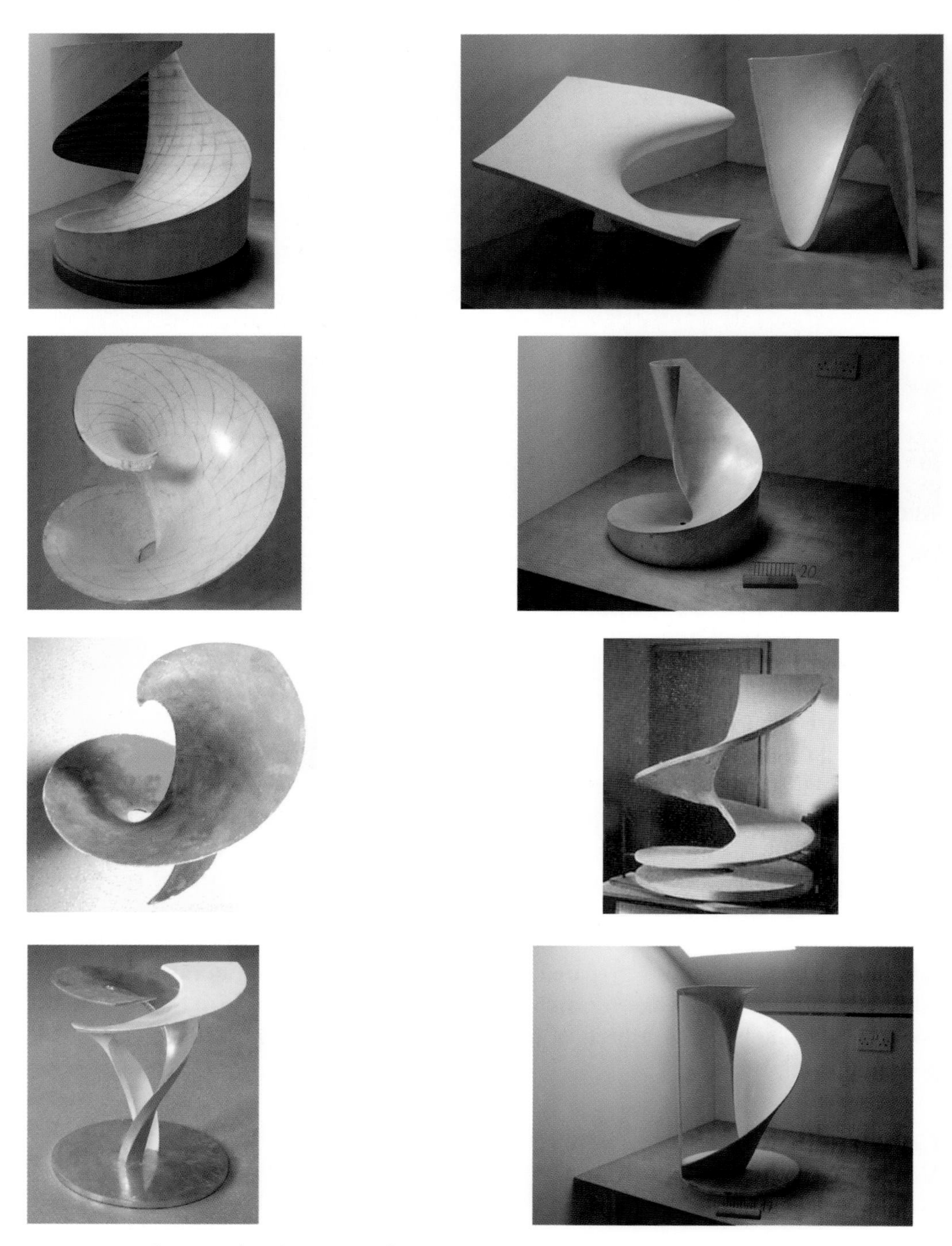

Fig. 23 A selection of path-curve surfaces constructed during the 1960s at the Institut für Strömungsforschung by John Wilkes for George Adams's research

Fig. 24 Rotational construction method used to make this type of surface

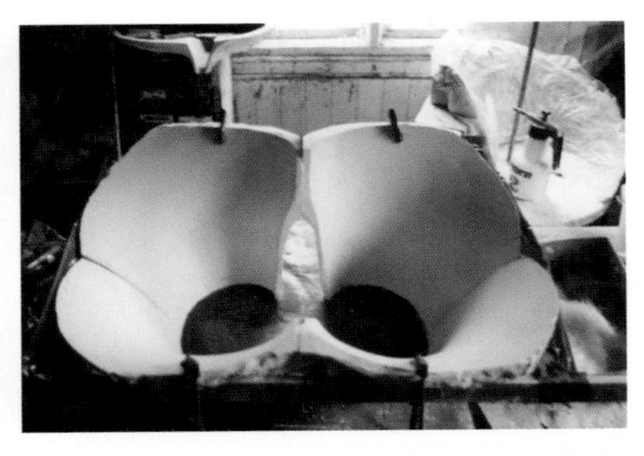

Fig. 25 Left- and right-handed sections of the same path-curve surface

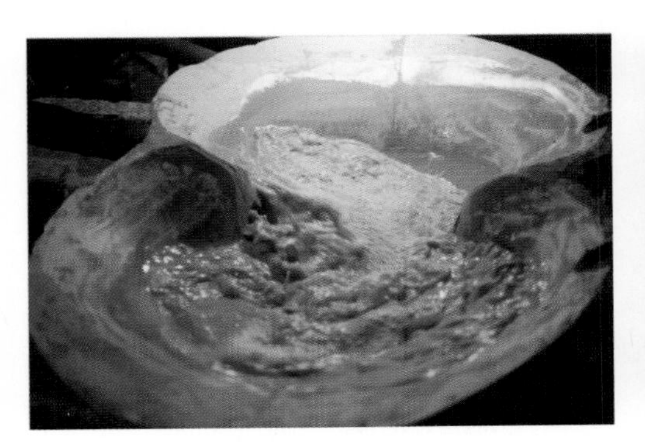

Fig. 26 Flowform vessel being tested with water for the correct function.

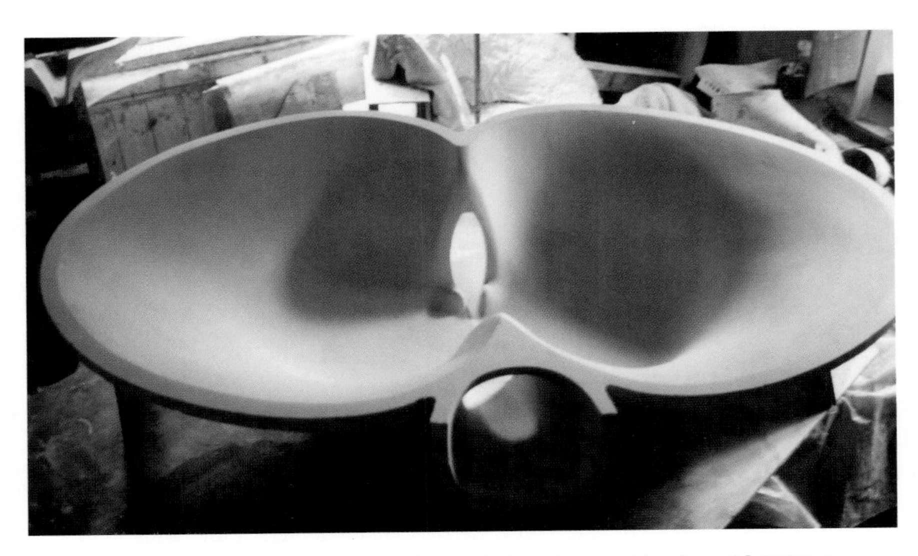

Fig. 27 Prototype of Flowform Vortex Cowhorn design (created in the mid-1990s)

Fig. 28 Cascade of two Flowform Vortex Cowhorn vessels, with water leaving vertically through two holes in the base of each form (March 2009)

Wheat seed germination comparing the Vortex and Vortex Cowhorn Flowform models

by John Wilkes, Nick Weidmann, Paul King and Jochen Schwuchow (2009–10)

Water was treated on 7, 13 and 14 October 2009 using both Flowform models. To determine qualitative influence of such designs treatment was continued for one hour so various tests could be carried out with different samples.

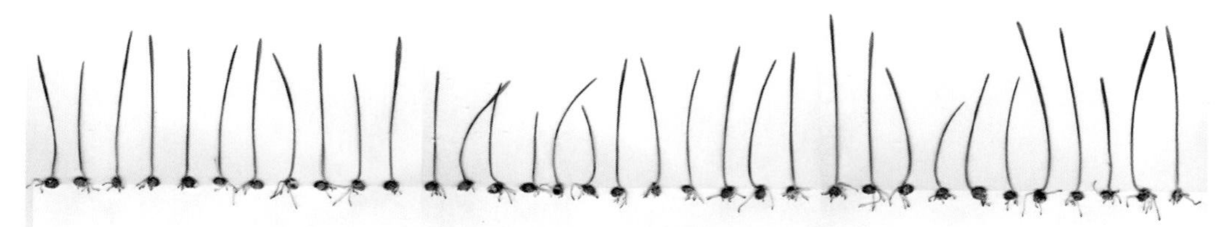

Fig. 29 Germinated from 16 October 2009: tap water control, one hour pump-oxygenated. Pressed 29 October

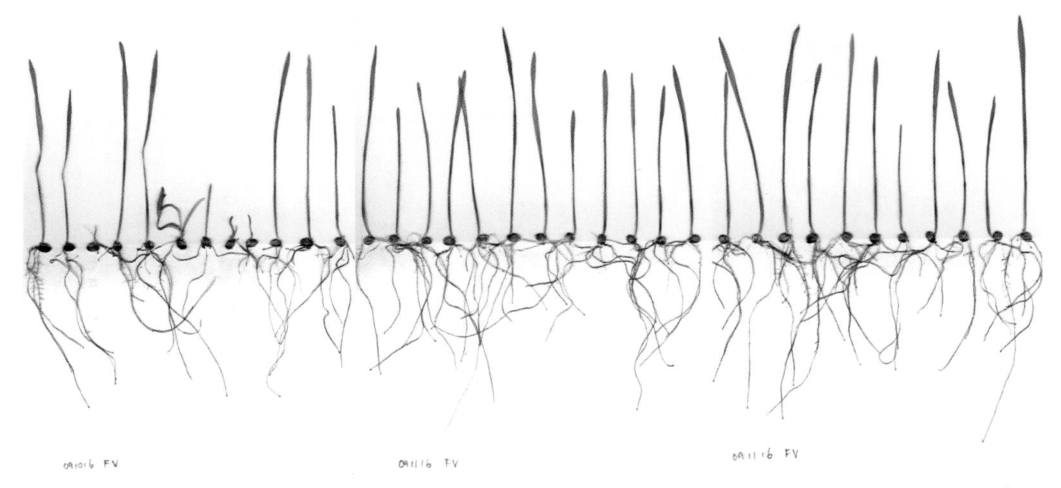

Fig. 30 Germinated from 16 October 2009: Vortex Flowform treatment for one hour. Pressed 29 October and finally mounted for comparisons and eventual measuring

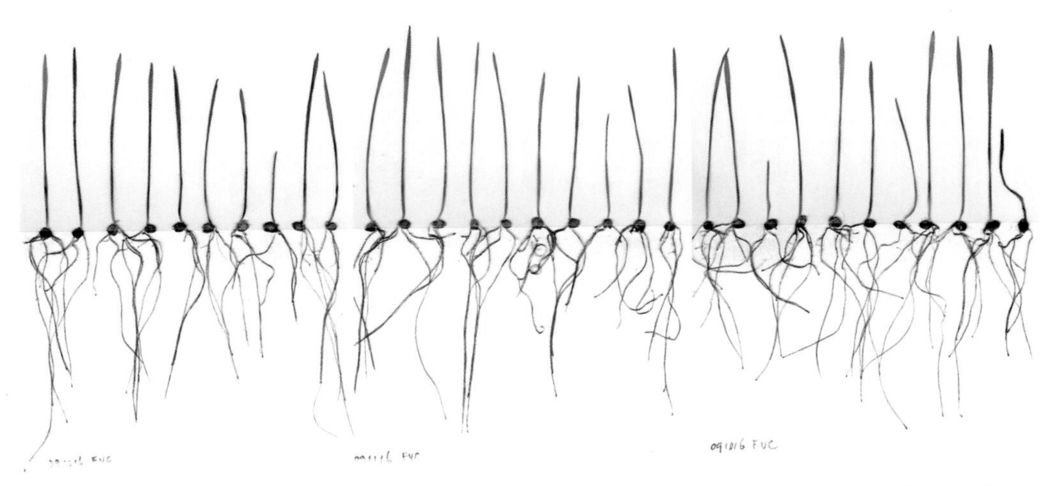

Fig. 31 Germinated from 16 October 2009: Vortex Cowhorn Flowform treatment for one hour. Pressed 29 October 2009

These preliminary trials indicated that water from the mathematically generated design, when used to germinate seeds, had a marginally stronger influence on root growth. The similarly oxygenated control had little response (see Figs 29, 30 and 31).

Samples were kept from water treated on 7 October and 13 October to compare with water treated on 14 October, initially establishing how long the water retains effects of treatment (see Figs 32 and 33).

Fig. 32 All glasses of wheat seedlings germinated from 14 October 2009. (Photographed 22 October)

Fig. 33 Wheat seedlings germinated from 14 October. Same seedlings as above (Fig. 32) photographed 27 October

Germinations started on 14 October with water treated on 13 October and 14 October are similar in root length. Water treated on 7 October produces root growth only a little less in length, showing a tendency to hold growth influence over that time period.

These experiments will continue to be increasingly rigorously investigated as we further develop our path-curve designs.

4. RESEARCH ON FLOWFORM EFFECTS

The whole water circulation has tremendous significance for the life of the earth. Just as the human organism could not live if it did not have its blood circulation, so the earth would be unable to live if it did not have its water circulation.

Rudolf Steiner (1861–1925)

Flowform phenomena and rhythm research

Rhythmic phenomena are fundamental to all life processes. In a Flowform cascade a multitude of rhythms can be generated with different frequencies and qualities of movement, dependent upon the surfaces over which the water is allowed to flow. Some results of the rhythm analysis work carried out so far indicate that not all the processes occurring are as yet understood. For example, some low-frequency components of the rhythm near the inlet appear to be of a magnitude comparable to the fundamental Flowform water pulse. We also need to understand other rhythmic components that differ in asymmetrical designs. But experience has shown that rhythmic patterns can be affected by surface and shape criteria.

Rhythms and cosmic influence

All organisms exist within a rhythmic context (periods of waking and sleeping, for instance), and all plants are subject not only to annual and seasonal but also to diurnal rhythms. Relating forms and rhythmic patterns to cosmic influences and rhythms is an important aspect of the work.

There is evidence that the effects created by rhythmical treatment depend on the time of day when it is applied (Hagel 1983), as well as upon the lunar cycle (Schikorr 1990, Loyter 2005, see p. 71 in the chapter on current research at the Healing Water Institute). Rhythmic treatment during an eclipse appears to have a growth-inhibiting effect on the water (Wilkes, personal communication). This suggests that Flowform water treatment is not only effective in isolation, but is acting as a kind of sense organ for cosmic influences. This needs further investigation over extended periods to consider the effects of other planetary rhythms.

Following the work of George Adams, Lawrence Edwards and Nick Thomas, the use of projective geometry for this purpose is being studied (Edwards 1982, 1993, Adams and Whicher 1979). Links to the work of investigators such as Viktor Schauberger may also prove fruitful. The use of mathematically designed Flowform surfaces, especially in connection with shape analysis, may also be viable as a result of these developments. Further progress in rhythm and shape analysis will depend on the theoretical insights gained in a mathematical and geometrical context.

Wheat and cress-growth experiments

The effects of Flowform water on plant growth have been tested experimentally and compared with control water that is oxygenated to the same degree but otherwise untreated. A number of systems have been established to conduct experiments under controlled conditions.

Wheat seeds are placed equidistantly on a mesh fixed inside a glass containing the water sample. The glass is placed on a turntable and rotated slowly for 10 days. The roots grow through the mesh into the water (Fig. 34a; also see Research Project 3 in the section on current research). The seedlings are photographed and the length of shoots and roots can be measured either manually or with a computer program.

Cress germination and growth experiments are carried out with a method developed by Hiscia Cancer Research Institute in Arlesheim, Switzerland. The chosen water sample is added to each envelope on a sheet of filter paper, cress seeds are lined up inside and the sealed envelope is hung in a light-tight box, where the seedlings are left to grow for four days, with humidity and temperature constant. After this period the results are scanned and records of the growth are measured with software developed for that purpose (Fig. 34b; for further details on the cress-growth method, see Research Project 1 in the section on current research).

Crystallization method

Crystallization with copper (II)-chloride was first developed by Dr Ehrenfried Pfeiffer – following a suggestion given by Rudolf Steiner – as a sensitive method to test specific biological substances, such as plant extracts and blood (Pfeiffer 1931,

Fig. 34 *(a) Wheat seeds germinating* on a mesh in a glass containing the water sample. *(b) Cress seeds germinating* on filter paper inside a plastic envelope

1935, Nickel 1968). Pfeiffer observed that the crystallization pattern of copper chloride, which as pure metal salt has a rather disordered structure, becomes strikingly coordinated and restructured under the influence of biological substances. By describing the texture, comparing various patterns and establishing specific types, these pictures can be evaluated morphologically (Fig. 35).

Examining different developmental stages in bean, radish, rape and cress, Magda Engqvist noticed that as the seedling develops the crystallization pattern becomes more branched and complex. In contrast, ageing processes result in a loss of

Fig. 35 *Crystallization dishes with Flowform water* (above left); Flowform water treated inside a geometric form (above right); tap-water control (bottom left); and crystallization chamber (bottom right). These are comparative examples used in test experiments

coordination in the typical crystallization pattern. Also environmental growth factors, such as light and shade, as well as various soil conditions (loamy, sandy, calcareous) modify the crystallization pattern with respect to richness of needle formation, branching of the bundles, and ratios between the inner, middle and outer zones of the picture (Engqvist 1970).

According to Selawry (1957), the crystallization images from plant extracts can be classified according to four fundamental types: root, shoot, leaf and flower, each showing a distinct crystallization pattern.

The importance has been highlighted of harvesting plants at the same time of day, preferably in the morning, since all plants are subject not only to annual and seasonal but also to diurnal rhythms (Selawry 1957).

Observations suggest that the more vigorous and alive the sample substance is, the more its formative influence *overcomes* the particular pattern of the copper chloride itself.

Copper chloride crystallization can be used as a sensitive, qualitative method to detect growth, ripening and decomposition processes in plant substances, to identify a range of environmental conditions, and also as a diagnostic tool for blood samples.

Capillary method

Following a suggestion by Rudolf Steiner in connection with their studies of plants, Eugen and Lily Kolisko at the Goetheanum Biological Institute developed the capillary method in the 1920s. They discovered that they needed to combine the action of metallic salts with the specific action of vegetal extracts to generate informative images (Kolisko and Kolisko 1978).

Capillary dynamolysis ('rising pictures'

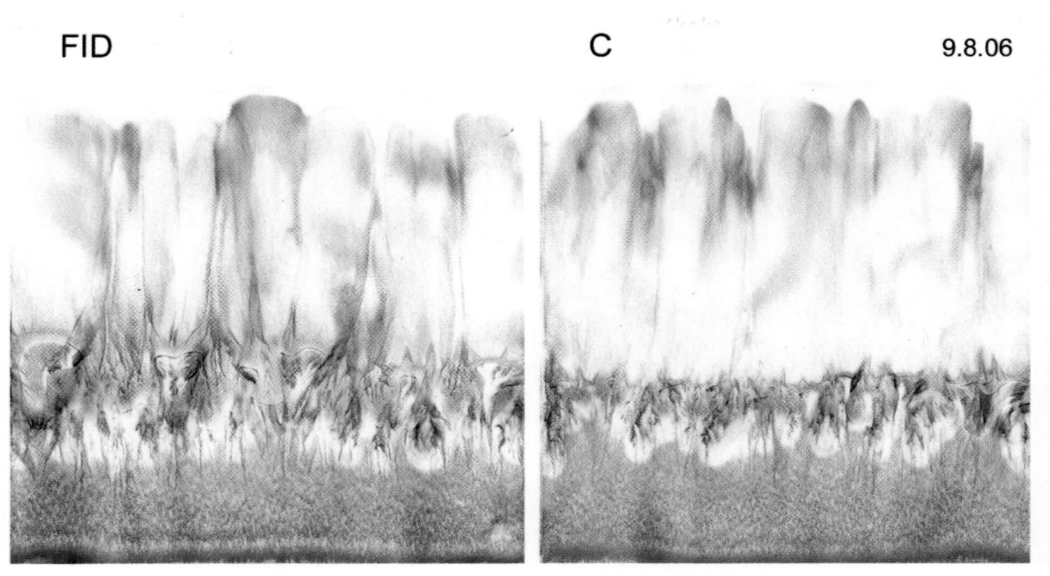

Fig. 36 ***Capillary method with Flowform water*** **rhythmically treated within a geometric envelope (FID), and untreated water as control (C). A number of replicates are normally used. (See also Fig. 61)**

method) is a chromatography method whereby a plant extract or water sample is allowed to rise vertically, by capillary action, through a special cylindrically shaped chromatography paper (Fig. 36).

By adding specific metal salts in dilution (silver nitrate and ferrous sulphate) which rise through the extract after it has dried, coloured patterns appear on the paper characteristic of the extract and the type of treatment. Capillary dynamolysis is thus known as one of the picture-forming or imaging evaluation methods. These patterns are read to reveal information about the properties and qualities of the extract or water sample (Barker 2005).

In blind experiments, Tingstad demonstrated the sensitivity of the capillary method by showing differences in the quality of carrots grown on the one hand with compost and biodynamic preparations and on the other with mineral fertilizers. None of the traditional quality investigations showed significant differences between the different fertilizer treatments (Tingstad 2002). The capillary method was also used to demonstrate the influence of moon phases and solar eclipses on plant saps (Engqvist 1977, Fyfe 1967).

Capillary and crystallization methods will be used routinely at the Healing Water Institute to complement quantitative analysis (Figs 35 and 36, also see Research Project 3 in chapter 5 on current research).

Round-filter chromatography (chroma method)

The microbiologist, biochemist and farmer Ehrenfried Pfeiffer spent several years of pioneer work and research developing 'round filter paper chromatography' into a practicable test method. This method is related to capillary dynamolysis, the difference being that patterns rise from the centre of a disc of filter paper placed horizontally into a glass or Petri dish (Fig. 37). Solutions of silver nitrate and sodium hydroxide containing the plant extract are allowed to rise with the help of a paper wick and spread out over the discs of filter paper.

Chromatographs in general can help to determine a need for further analysis, and can assist in obtaining a more complete picture of soil qualities. However, interpreting chromatographs requires experience and a sufficient supply of controls for comparison. Each type of soil needs its own standard series, just as plant/food chromatographs require a standard series for each plant or food type.

Drop-picture method

The patterns that arise when drops of a liquid fall into a larger quantity of the same or another liquid have long been of interest. One such form is the vortex ring, which was first described by Rogers in 1858 (see Smith 1974). These rings appear as the drop strikes the surface of a liquid and begin to descend (easily made visible with droplets of ink descending in water). After a short period, the descending ring becomes unstable and splits into a number of smaller rings, which in turn may split up as well. Depending on the falling speed, drop size, temperature and other parameters, other forms may arise which strongly resemble organic life forms (Hatschek 1919, D'Arcy Thompson 1942).

A striking result of these findings is the

Fig. 37 ***Experimental set-up for round-filter chromatography*** **(by Dr Uwe Gier at the Goetheanum Lab, Dornach, Switzerland)**

high sensitivity of the liquid forms to parameters such as density (Hatschek 1919), fall height of the drops (Chapman and Critchlow 1967, Thomson and Newall 1885), viscosity (Hatschek 1919, Thomson and Newall 1885), surface tension and others. This sensitivity of interfaces, and thus of the associated patterns and forms which they enclose, to the physical conditions of the system has been utilized as a means of testing the quality of drinking water and water in general (Schwenk 1967).

The drop-picture method was developed by Theodor Schwenk, based on experiments done by L. Kolisko. The parameters involved, such as drop height, layer depth, temperature and viscosity, have been studied in some detail ever since (for example Jahnke 1993, Schwenk 1967, 2001, Smith 1974, 1975, Wilkens et al. 2005).

In the drop-picture method, the sample liquid, such as a water sample or plant extract, is mixed with glycerol to form a thin layer in a glass dish. Pure distilled water is then allowed to drop into the glycerol/water mixture at a frequency of approximately 1 drop per 5 seconds. The resulting patterns are made visible as density gradients and can be correlated with the quality of water (Fig. 38). According to Schwenk, this method can also be used for testing pharmaceutical preparations as well as mineral springs, for examining the influence of different materials coming into contact with water, and for tracing the course of pollution in rivers (Schwenk 1967).

At the Healing Water Institute we are currently setting up the drop-picture method, potentially to help determine the effects of Flowform rhythmic movement.

Quality and properties of Flowform water

Research ongoing for many years has examined to what extent Flowform-induced rhythms and figure-of-eight flow patterns change the qualities of the water stream that runs through them. Apart from their aesthetic qualities, Flowform cascades do appear to have significant ecological and environmental applications and effects.

A comparative study undertaken by physicist Christian Schönberger and Professor Christian Liess in Überlingen, Germany, of research articles on Flowform technology indicates that the qualities and properties of water running through them are altered (Schönberger and Liess 1995). Permeated by rhythmic movements, Flowform-treated water not only becomes highly oxygenated but also more intensively supports regenerative rhythmic and biological processes (Wilkes 2003).

Water quality is generally tested for its chemical condition and this is seen as the main quality issue by western science,

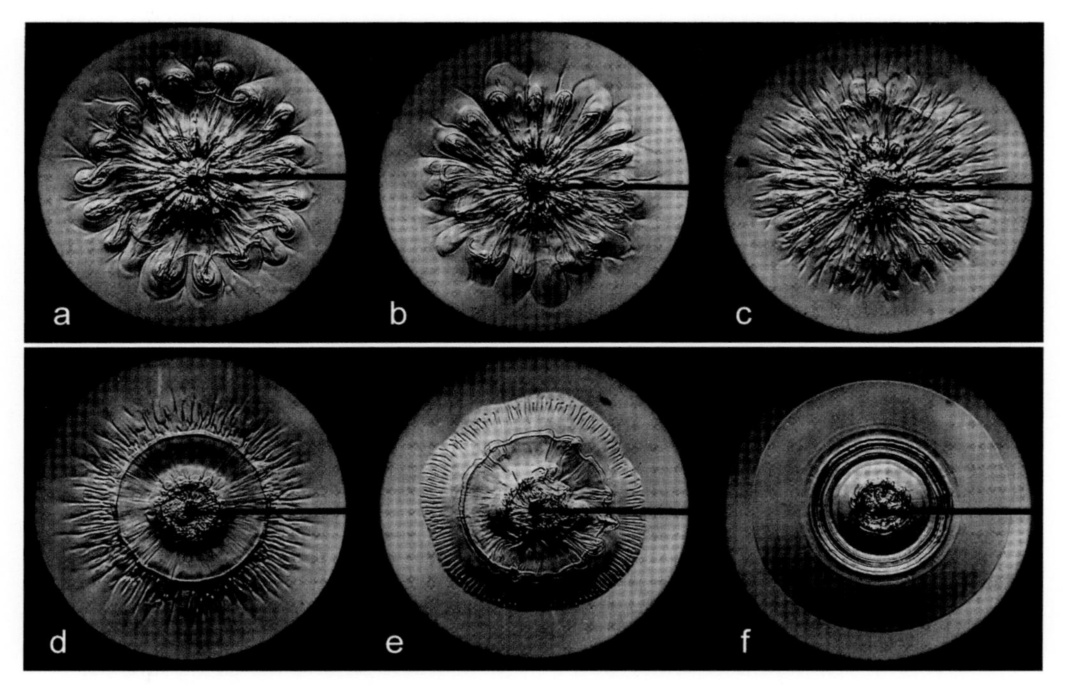

Fig. 38 *Typical patterns arising in drop-picture images:* (*a*) variform, (*b*) single-leafed, (*c*) raylike, (*d*) radiating outwards, (*e*) radiating inwards, (*f*) disclike (W. Schwenk)

along with water's organic content. However, modern scientific research is reopening the concept of energetic quality in the context of material physics studies of informational frequency.

It is precisely this concept of life-supporting quality, alongside water's chemical and organic content, that has been the long-term focus of the Healing Water Institute and its antecedents going back to the 1920s.

Oxygenation, organic content and pH

The effectiveness of a system of wastewater treatment is frequently measured using the following parameters:

- **Dissolved oxygen.** In water with high levels of organic compounds it is likely that dissolved oxygen will be depressed (Tebbut 1992). For biochemical oxidation to occur there must be sufficient aeration that can be measured by the concentration of dissolved oxygen present in the water.
- **Biochemical oxygen demand (BOD).** This can be used as an indication of the concentration of organic compounds in water (Tebbut 1992). By measuring the change in biochemical oxygen demand it is possible to determine if the amount of organic material is being reduced.
- **Faecal coliforms.** These are an indicator of bacterial and other pathogenic contamination.

In 'The Channon' in New South Wales, Australia, four test runs in a Flowform system receiving communal laundry water showed the *biochemical oxygen demand* (BOD) was reduced from an average of 424 to less than 20 mg/l over a 105-minute period. Within this period, *faecal coliform* counts were reduced from an average of approximately 3100 cfu/100 ml to a level of 500 cfu/100 ml, and the amount of *dissolved oxygen* increased steadily from 0.1 to 3.9 ppm. The dissolved oxygen (DO) concentration for 100% air-saturated water at sea level is 8.6 mg O_2/l (ppm) at 25°C and increases to 14.6 mg O_2/l (ppm) at 0°C. These results demonstrate the capacity of the Flowform system to break down organic matter and reduce bacterial contamination (Spencer 1995).

In two sample tests conducted in New Zealand at the Fuglistaller farm by the Taranaki Regional Water Board (1989), over a 5-day period using the 50 litre per minute Flowform Järna series to treat a 25 cubic metre dairy shed effluent pond at night-time, the BOD levels only dropped from 280 g/m^3 to 88 g/m^3. To be sprayed out onto paddocks as liquid fertilizer, 30 g/m^3 would be needed; but as rain arrived in day 6 we stopped the trial. It is likely that the BOD would have continued to drop (Trousdell EDRI Report 1990).

In Solborg, Norway, a system of ponds for wastewater treatment using Flowform vessels improved oxygen enrichment from 30% to 90%, as detected between the inlet and outlet within the Flowform cascade (Mæhlum 1991, Schönberger and Liess 1995). The continuous rhythmic movements induced by the cascade also prevented freezing of the pond in winter.

In the same installation, between the deposition pit and the 3rd pond, the content of chloride in the water was reduced by more than half (Mæhlum 1991, Schönberger and Liess 1995). Since chloride scarcely breaks down in nature, it was assumed that chloride reduction was caused by intrusion of surrounding water and dilution. Measurements of the inflowing and outflowing water however showed that the chloride reduction was greater than would be possible through dilution (Mæhlum 1991).

In trials in Holland, the transmission coefficients for oxygen in a Flowform cascade were very similar to a meandering step cascade (Järna design 0.39/Malmö design 0.45/Olympia design 0.49/step cascade 0.46) (De Jonge 1982, Schönberger and Liess 1995).

Flowform water had a pH-value that increased by 0.77 compared to untreated water, whereas the electrical conductivity decreased (Brückmann et al. 1992).

In New Zealand, oxygenation research was conducted in 1987 and 1988 to assess how different Flowform design types affected water. The *Beehive, Järna* and *Taruna* models were compared.

The Hawkes Bay Regional Council water board scientist and Rob Dewdney carried out both field and laboratory tests using a dissolved oxygen meter and the Winkler test.

Bore water at 13°C with high levels of calcium was run once only through a Flowform cascade with 12 *Beehive* models, which are 4- and 5-chambered 'lung' designs with extra vortical chambers. This method was repeated 20 times to gain average readings. The oxygen readings taken at the start averaged 1.4 parts per million (ppm) and at the end 7.0 ppm, showing an increase of 5.6 ppm over a distance of 4.8 metres. Each unit increased dissolved oxygen by an average of 0.465 ppm.

Similar testing with the Flowform *Järna* model, which is a 'kidney' shape emphasizing mixing and restructuring, showed an average increase per unit of 0.2 ppm.

This was repeated six months later in 1988, but with a cascade of 23 of the same *Beehive* models with the same bore water, giving a start reading of 0.7 ppm, a middle reading of 6.4 ppm and the end reading of 8.95 ppm. The water was run through this cascade once only, but again repeatedly with new water to gain average readings. Increase in the last 12 vessels was 0.163 ppm per unit. Oxygenation in water above 9 ppm becomes saturated and is supersaturated above 11 ppm.

In order to find out more about Flowform capacity to oxygenate at the (super)-saturated dissolved oxygen end-range, pre-oxygenated tap water and bore water was run through a 'heart-lung' *Taruna* cascade, at first just run through once to the 7th unit, and then reticulated repeatedly using a centrifugal pump.

The start reading was 8.1 ppm; and run once through to the 7th unit, the readings then showed 10.8 ppm. Running a total volume of 800 litres of water for six minutes (at 100 litres per minute) through the cascade gave a reading of 12 ppm, which is supersaturated. After three minutes, readings were 11.4 ppm. Natural oxygenation in the local Tukituki River was tested, reading 12 ppm after 40 metres of gentle rapids, and 7.6 ppm before it. (I. Trousdell 1989.)

This demonstrates the need for more research into oxygenation, especially as there are some indications that oxygen introduced into water through Flowform activation may stay longer than through the method of spraying the water up into the air (Chris Weeden, personal communication).

At Laverstoke Park Farm soil testing laboratory in England testing was done in 2009 to compare the dissolved oxygen (DO) content of water treated with the Flowform *Vortex* model and conventional air-pump oxygenated water.

Although both systems aerate water satisfactorily, once the pumps are turned off, the Flowform *Vortex*-treated water has an ability to maintain its DO levels compared to the bubble diffuser method, which immediately and rapidly declines within minutes. This preliminary work suggests that the Flowform *Vortex* design will hold DO levels in water for up to 8 days (which is when the experiment stopped, not when the DO levels dropped!).

The ability of the Flowform *Vortex* model to sustain microbiological growth in liquid compost extracts was assessed in the light of these oxygenation results, and once again it performed better for the same period of up to 8 days, compared to conventional aeration methods, not only during the aeration process but also after dynamization ceased. Microbial activity would normally significantly decline once conventional methods of aeration ceased. (J. Williams, 2009.)

Flowform treatment appears to be a more efficient and durable way to achieve oxygen saturation, due to the more natural folding and weaving movements of the water which enrich its internal microstructure.

Density, temperature, viscosity and flow rate

The influence of parameters of density, temperature, viscosity and flow rate on water rhythms was studied by Martin Strid in a Flowform cascade with four vessels at the University of Luleå in Sweden. The water viscosity was altered with polyeteneoxide powder in the range of 10^{-6} to 10^{-2} m^2/s; water density was altered between 1000 and 1170 kg/m^3 by addition of salt, and temperature ranged from 5° to 48°C.

The frequency of the water pulse in a Flowform vessel did not depend on density, viscosity or temperature of the fluid, but was found to depend only on the quantity of water.

In one particular Flowform vessel, for example, the pulse started at a flow rate of 3.0–3.6 l/min with a frequency of 104.0 min^{-1} (1.73 Hz), whereas at 7 l/min the frequency increased to 107.6 min^{-1} (1.79 Hz). At different positions within the Flowform vessel, electrodes were immersed in the flowing water in order to measure electromagnetic properties. It was found that the voltage in the vessels pulsated with the same rhythm as the water (Strid 1984, Schönberger and Liess 1995). This enables rhythm frequencies to be measured accurately.

Rhythms and sound frequencies

Nick Thomas investigated rhythms in single and multiple Flowform cascades using Fourier analysis (Thomas 1983). He measured the depth of the water electrochemically, and recorded the measurements at regular intervals to obtain a frequency plot of the rhythm.

The results obtained showed a high diversity of frequencies in different designs, some displaying very strong secondary frequencies.

Rhythms were often complex even in single vessels, showing both variation of the fundamental frequency (corresponding to the visible pulsing of the water in the form) and distinct frequency components (Thomas 1983). Fundamental frequencies were in the range of 0.68–0.36 Hz, corresponding to a period of 1.5–2.8 s for the pulse in different Flowform types. Some models exhibited a high frequency component typically around 12 Hz, which might be attributed to surface waves.

Sound frequencies were also measured on a *Sevenfold* Flowform cascade. The sound in this cascade was found to have the character of a small, turbulent stream due to the wide spectrum of rhythms. Frequency spectra in all individual Flowform designs have relatively broad peaks in the 1000–1600 Hz region. The results indicate that the major sound-generating mechanism is the same for all designs (Kristiansen et al. 1993).

The sound in a Flowform cascade or a natural brook is thought to be caused by falling water catching and taking down little bubbles of air. Sound is generated when these bubbles burst, much the same as the sound heard when a drop of water impacts on a water surface.

The sound-producing part of flowing water is therefore characterized by whitish, turbulent water. The darker and less turbulent parts, where water gently slides to the pool below, do not produce any sound.

Flowform effects on plant growth and morphology

Lunar and planetary influences on plant growth

There is some evidence that the effects on plant growth created by rhythmic water-treatment depend on the time of day when it is applied to the water (Hagel 1983), as well as upon the lunar cycle (Schikorr 1990, Loyter 2005) (see Research Project 1 in the section on current research).

A number of researchers have established that the different phases of the moon and planets have a considerable effect on the germination and growth of plants (Edwards 1993, Endres and Schad 2002, Fyfe 1967, Schwenk 1967, Thun 2003, Thun and Thun 2004).

Plants appear to respond to moon phases more sensitively after being treated with Flowform water (Schikorr 1990, Loyter 2005, see Research Project 1 in chapter 5). The biggest enhancements of root length with Flowform water occur at the dates of new moon sowings, whereas full moon sowings result in minimal enhancement, and the moon in perigee results in a slight inhibition (Schikorr 1990, Schönberger and Liess 1995).

These experiments indicate that Flowform treatment might well amplify the water's sensitivity to growth-promoting or growth-inhibiting influences.

Generally, these results confirm the assumption that Flowform water treatment technology is not only effective in isolation, but is acting as a kind of sense organ for cosmic influences. This is indeed its main function: to absorb and mediate these influences on the living organism and thus embed it in its environment. No living thing can exist in isolation.

Influence on plant germination, weight and length

The rate of germination of wheat was demonstrated to increase by 11% in Flowform-treated water compared to untreated water (Hoesch et al. 1992).

When radish was exposed to Flowform water, the fresh weight of the bulb was higher compared to untreated water (Hagel 1983). Differences were most significant after treatment in the early morning, while the effect appeared to decrease towards the afternoon.

Experiments with wheat (Schikorr 1990) and cress (Loyter 2005, see Research Project 1 in chapter 5) carried out at Emerson College, Sussex, indicate that plants treated with Flowform water had greater average root lengths compared to controls grown with water aerated with a pump instead.

Influences on plant phenotype

Research in connection with biological water purification was conducted at the Warmonderhof Agricultural College in Holland with water that passed through a Flowform cascade and a non-rhythmic step cascade (Fig. 39).

Over a period of four years, the flora growing in the ponds downstream from the two cascades developed in different ways.

The Flowform-treated water stimulated generative growth and phenotypes characteristic of plants grown in the light. Flower development was pronounced, flowering occurred earlier, with stems more upright and smaller leaf production, and colouring was deeper in autumn.

In contrast, the step cascade induced vegetative growth typical of plants grown in the shade. Here leaves were wider and flower development was less pronounced (Van Mansfeld 1986, Wagenaar 1984).

Macrofauna and microbiological effects

In the step cascade system in Warmonderhof mentioned above, a higher number of species preferring darker habitats appeared (deep-water and bottom dwellers). These have softer and rounded shapes, move slower and often go through a life cycle with a flying stage, such as midge larvae.

In the Flowform system, more organisms were found that prefer regions of light (upper water layers and surface); these, such as water mites and crustaceans, have more differentiated and indented forms, quicker, more nervous movements, and go through a life cycle that remains in water.

The mobility of fish was higher in the Flowform system compared to the step cascade (Van Mansfeld 1986, Wagenaar 1984, Schönberger and Liess 1995).

Observations showed that in the step cascade system the water appeared cloudier and had a musty smell of ammonia, whereas the Flowform system had clearer water and a smell like humus or hay (Van Mansfeld 1986).

In a system for wastewater treatment in Järna, Sweden, containing seven Flowform

Fig. 39 *Biological water purification system* at Warmonderhof Agricultural College in Holland: (*a*) step cascade and Olympia model seen from above, (*b*) Step cascade (left), Olympia (centre) and Malmö model (right) cascades

cascades it was shown that pond water could be efficiently cleansed with respect to pathogen bacteria (Sernbo and Fredlund 1991, Schönberger and Liess 1995).

The installation outlet flows into the Järna fjord (Baltic), where the treated wastewater is of such high quality that micro-organisms and higher organisms prosper as well as they do in untreated fjord water (Alleslev 1987).

Flowform stirring of biodynamic preparations

Research conducted at Emerson College in Forest Row, England showed that Flowform stirring of biodynamic preparations (Fig. 40, right) has similar effects to hand stirring, and resulted in an increase of the yield of wheat by 22–25%, whereas machine stirring improved the grain yield by only 11% compared to control plants (Schikorr 1994, Schönberger and Liess 1995).

Observations made over a 20-year period on numerous farms in New Zealand that only used Flowform stirring of their biodynamic preparations showed that the number of earthworms increased, animals were healthier, and the soil was darker and more loose and friable (Trousdell and Proctor 1991, Schönberger and Liess 1995).

The New Zealand and Australian national biodynamic associations accept Flowform-stirred preparations in their Demeter standards because of extensive success in using them.

Researcher Dr Walter Goldstein reports from the Michael Fields Agricultural Institute at East Troy, Wisconsin, that the Flowform Vortex model (Fig. 40, right) using an Archimedean screw (Fig. 20, left) as a water re-circulation device produced preparations equivalent in quality to good hand stirring.

Using a conventional centrifugal pump with the same Vortex model resulted in somewhat lesser quality, whereas the use

Fig. 40 ***Biodynamic preparation stirring*** **using the Vortex model, which has two vortex holes that develop deep left and right vortices that fall into repeating chaos chambers. Note the application by aeroplane, AUS (left).**

of a mechanical stirring machine achieved a lower quality result (personal communication).

Biodynamic food production and Flowform effects

At Herzberger bakery in Fulda, Germany, the effect of Flowform-treated and untreated water used for baking bread was examined. The treatment consists of running the water over a number of granite slabs followed by a Flowform cascade (Fig. 41).

It was found that the amount of water uptake in the bread is higher after treatment and therefore the bread remains fresh and free of mould for at least two days more than that made with untreated water. Besides, the volume of dough is increased by 4%, while the consistency and taste is significantly improved (Strube and Stolz 1999, Brückmann 1992, Schönberger and Liess 1995).

Preliminary indications of other Flowform effects

Experiments with wheat seedlings also gave some indications of a longer-lasting sensitizing effect of Flowform treatment (Wilkes, personal communication). Further experiments over extended periods are needed to confirm these results.

Research by Peter Alspach, MSc, of the New Zealand Ministry of Agriculture and Fisheries in the mid-1980s showed Flowform water tended to precipitate inorganic

Fig. 41 ***At Herzberger bakery in Fulda*, Germany, water flowing over granite slabs (left) followed by a Flowform cascade (right)**

iron out of water, equivalent to the results of the influence of magnets.

Rhythmic treatment and electromagnetic properties of water

In conventional wastewater treatment, water is chemically cleaned, and cleansed of bacteria and pollutants such as lead, cadmium and nitrate.

Research indicates, however, that after processing in a conventional sewage plant water still contains certain electromagnetic frequencies that can be harmful (Ludwig 1991). Certain electromagnetic frequencies of water polluted with heavy metals have been detected in cancerous tissue, such as the frequency 1.8 Hertz (Gross 2000b). This frequency was detected in the drinking water of a German capital, even after being distilled twice (Gross 2000b, Ludwig 1991).

Thus it can be concluded that after any cleansing – purification, chemical treatment, filtration or even distillation – the pollutants' harmful information in the form of electromagnetic frequencies/oscillations can be transferred to the human organism. They are measurably present in the water molecules both before and after conventional treatments.

So even when our fresh water is chemically purified it is still physically charged with pollutant information, both before and after any conventional treatment. It is not the chemical substance that affects the human organism when we drink this water, but the undesirable frequencies.

Ludwig reports that repeated vigorous vortical treatments are the best way to neutralize undesirable information remaining in water after removal of physical pollutants (Ludwig 1991).

This points towards the possibility of cleansing water of unwanted electromagnetic frequencies by treating it with Flowform eco-technology, which allows water to move vortically in repeating flow patterns.

5. RECENT RESEARCH AT THE HEALING WATER INSTITUTE

In an age when man has forgotten his origins and is blind even to his most essential needs for survival, water along with other resources has become the victim of his indifference.

Rachel Carson (1907–64)

Numerous colleagues around the world have carried out research since the discovery of the Flowform principle by John Wilkes in 1970. The main research however has been conducted at the UK Healing Water Institute, based at Emerson College, Sussex, England.

One of the main areas of ongoing research is investigating and analysing the effect of Flowform treatment on plant growth. Various research projects were carried out at the Healing Water Institute using lettuce plants in the field (RP2), as well as cress (RP1) and wheat (RP3) under laboratory conditions.

Research Project 1: The effect of Flowform treatment on cress germination and growth

by Orit Loyter (2005)

Abstract

Investigations into the effect of Flowform-treated water on cress germination and growth, in comparison to tap water, were carried out from May to August 2004. The tests were performed eight times a month in relation to the moon cycle. After a 4-day growth period the length of shoots and roots was measured as an indicator of water quality. It was found that on average the Flowform treatment promoted growth, with the largest difference being observed in root growth. The experiments indicated that seedling growth might show a pattern according with the lunar cycle.

Introduction

This research project aimed to examine the effect of Flowform-treated water on plant growth. The first stages of cress germination and growth were examined, as it is known that during initial stages of growth the plant is very sensitive to its surroundings; previous research was checked for relevance.

Maria Thun (2003) and others showed that treatment during those first developmental stages had an influence on the whole life of the plant. One of the most extended experiments examining the influence of Flowform-treated water on plant growth was conducted in Holland by Prof. Jan Dick Van Mansfeld of Warmonderhof Farm School (Van Mansfeld 1986). Other research examined the effects of Flowform eco-technology as part of a wastewater treatment system, for example Spencer (1995) and Mæhlum (1991).

In his agricultural lectures, Rudolf Steiner referred to the close connection between plant growth and planetary movements (Steiner 1938). These ideas have been incorporated into biodynamic agriculture. A number of researchers have established that the different phases of the moon, as well as other planetary influences, have a strong effect on the growth of plants (Lawrence Edwards 1993, Georg Schmidt 1984, Theodor Schwenk 1967, Maria Thun 2003).

Research by Freya Schikorr (1990) suggested that it is possible to enhance lunar influences on plant growth by treating water rhythmically in Flowform cascades.

It was found that Flowform treatment promoted root growth especially, and that the rhythmic movement of the water seems to have a positive effect on biological processes resulting from its enhanced sensitivity.

Materials and methods

The cress-growth experiments were carried out using a method developed by Hiscia Cancer Research Institute in Arlesheim, Switzerland. A sheet of Whatman no. 1 filter paper (14 × 8.5 cm) was placed in a plastic envelope (21 × 12.5 cm) and sealed for the duration of growth. To each envelope 3 ml of the specific water sample were added.

Twelve biodynamic cress seeds (from Botton Village, Yorkshire, biodynamic seed producers, and, for some experiments, from Stefan Baumgartner in Hiscia) were lined up in a row equidistantly inside the envelope. The seeds were placed horizontally across the filter paper 10 cm from the bottom using a cardboard pocket into which the envelope was temporarily placed. Six replicates, each in a separate envelope, were prepared from each water sample. After closing the seal the envelope was hung in a light-tight box on two rods. The seedlings were left to grow in the dark for 96 hours at constant humidity and temperature (approx. 55% and 21°C).

Following an initial period of studying the method and its applications, 22 experiments were performed between 4 May and 31 July 2004.

Water treatment

Four different water samples were used in each experiment: the control and three out of four available Flowform cascades. All cascades were made from a series of Flowform laboratory stackable models.

Three Flowform cascades (see Fig. 42 below, FG, FJ and FJI) were used, consisting of 16 stackable plastic Flowform units each. Cascades FG and FJ were placed in different locations and cascade FJI was partially surrounded by a geometrical form – a projected icosahedron made of bronze rods. The fourth cascade (FW) consisted of 38 units of the same laboratory stackable models.

The projected icosahedron was developed by John Wilkes using projective geometry methods. A similar projected geometric envelope surrounding a vertical Flowform cascade was set up in Norway in a factory that produced grain milk. Previous studies by John Wilkes suggested that such a form could enhance the Flowform effect on water.

The water was circulated through Flowform cascades FG, FJ and FJI for a period of one hour by a centrifugal pump ((Kockney Koi Yamitsu submersible fountain pump, model KKYFP2400) at a flow rate of

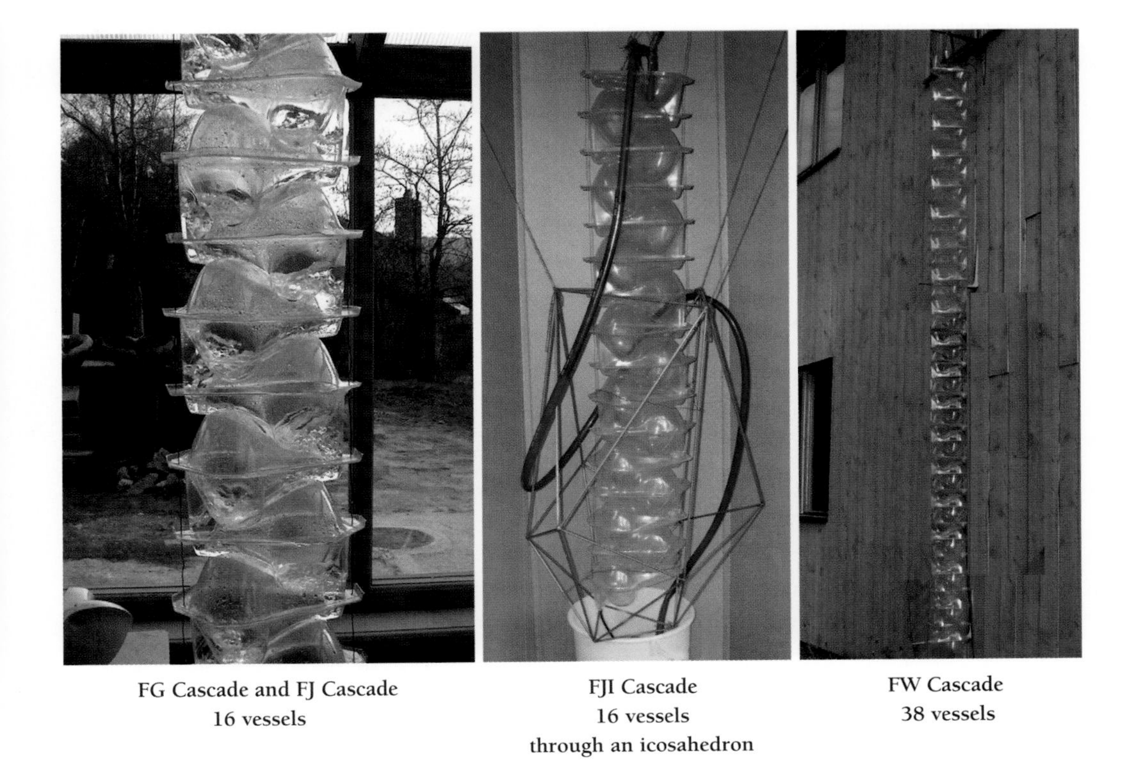

FG Cascade and FJ Cascade
16 vessels

FJI Cascade
16 vessels
through an icosahedron

FW Cascade
38 vessels

Fig. 42 The stackable Flowform cascades used in the experiments

6 l/min prior to taking a sample at the outlet of the last Flowform vessel.

Water flowed into cascade FW directly from the tap through a hose, passing through once, whereas electrical pumps were used on the three other cascades. All water samples were taken at 8 a.m.

To summarize: the following cascades were used for the experiments (see Fig. 42 above):

FW cascade consisting of 38 stackable Flowform vessels through which the water is allowed to flow once.

FG and FJ cascades consisting of the same 16 stackable vessels but placed in two different locations.

FJI cascade, the same as FG, but partially surrounded by a projected icosahedron made of bronze rods. The projected geometrical form was 100 cm and the cascade 150 cm high.

Measurement and statistical analysis

After the 4-day growth period the results were scanned for each sample and the length of shoots and roots was measured using National Institute of Health (NIH) image software and a special macro program (designed by J. Schwuchow). The program makes it possible to digitize shoot and root segments and calculate the total lengths.

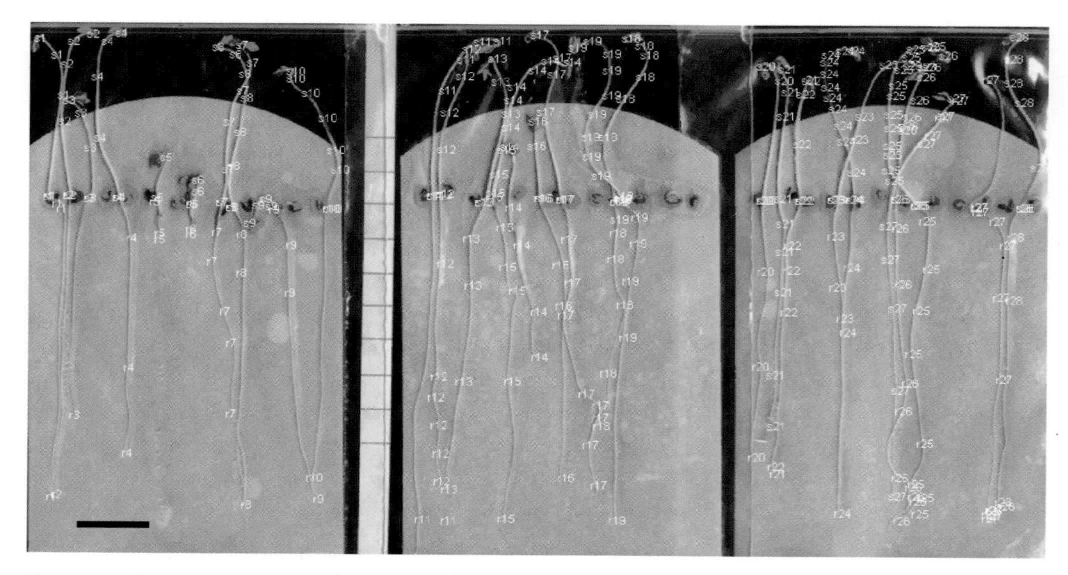

Fig. 43 Three envelopes with cress seedlings show overlaid size markers using NIH image software to demonstrate the measurement technique. Bar = 20 mm

In each of the experiments, the length of shoot, root and total growth were indicated in either absolute values or in percentages of growth with respect to the control. In all experiments t-tests were carried out, and results were considered significant where P is less than 0.1. Mean growth was calculated using only the germinated seeds.

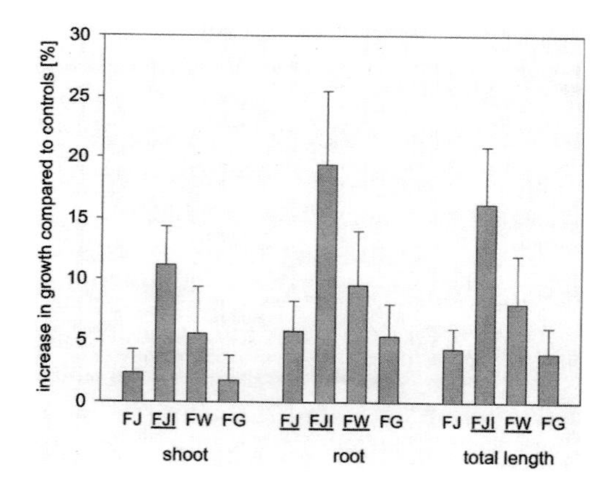

Fig. 44 *Mean increase in cress-seedling growth treated with water from Flowform cascades* FJ, FJI, FW and FG with respect to the control. The mean results of all 22 experiments are shown for shoot, root and total seedling length. Samples that are statistically different from controls are underlined. Error bars: standard error [SE]

Results and discussion

Seedling growth in Flowform water

The mean increase in the shoot, root and total seedling length compared to the growth of the control is shown in Fig. 44. The averages were calculated by using the germinated seeds only. The average germination rate was about 60–70%.

To check the statistical significance, t-tests were carried out and the parameter P with $0 \leq P \leq 1$ was determined. The closer

P gets to 0, the more significantly different the two sets of data are. When $P \leq 0.1$, the groups are considered statistically significantly different (underlined labels in Fig. 31).

The increase in root growth of plants treated with Flowform water compared to untreated controls was larger than the increase in shoot growth. Mean increase in shoot growth from all Flowform water is 4.4 ± 1.3% [SE] with respect to the controls, compared to a 8.8 ± 1.9% [SE, statistically different] increase for root lengths.

Comparing the results from different Flowform-treated water demonstrated that cascades FJI and FW showed a much larger additional growth (11.2 and 5.6% respectively for shoots, and 19.4% and 9.5% for roots) in comparison with cascades FJ and FG (2.4% and 1.8% for shoots and 5.8% and 5.4% respectively for roots).

The highest mean increase in root growth was found after treatment with water from the FJI cascade (19.4 ± 6.1% [SE], Fig. 44).

In a few experiments, the amount of oxygen in the control water was increased, which did not result in any detectable effect. The location of FG had no significant effect, since cascades FG and FJ produced similar results.

Patterns of root growth in May, June and July in relation to the moon phases

Possible correlation of patterns of variation in root growth with moon phases was investigated. For this purpose, all root lengths from experiments in the months of May, June and July 2004 were plotted chronologically and compared.

It appeared that there may be a correlation between root growth and the moon phases. In all three months, root lengths reached a maximum about one week after full moon and, in some cases, a somewhat smaller peak a few days after new moon.

Root lengths were at a minimum at the time of or shortly after full moon and at or shortly after new moon.

It can be noted that in all samples in May, June and July (Fig. 45 a, b and c) the gradients of changing root lengths were fairly similar, i.e. the connecting lines show a similar gradient.

This indicated that there might be some changing effect on growth during the course of the month, which may influence all the samples in a similar way.

This pattern was more obvious in June and July (Fig. 45 b, c) than in May (Fig. 45 a).

To examine whether Flowform technology had a specific influence independent of general fluctuations during the course of the month, the average root growth of each Flowform-treated seed sample was divided by the control growth in the same experiment.

This data indicates the relative growth (%) of samples treated with Flowform water with respect to the controls, thus showing the specific influence.

At certain times during the course of the month, Flowform water appeared to promote growth significantly (around 24 May, 2 and 27 July), while at other times there was no effect (around 11 May and 16 July), or possibly even a slight inhibition of growth (19 May and 17 June). In all three months there was a minimum of growth in

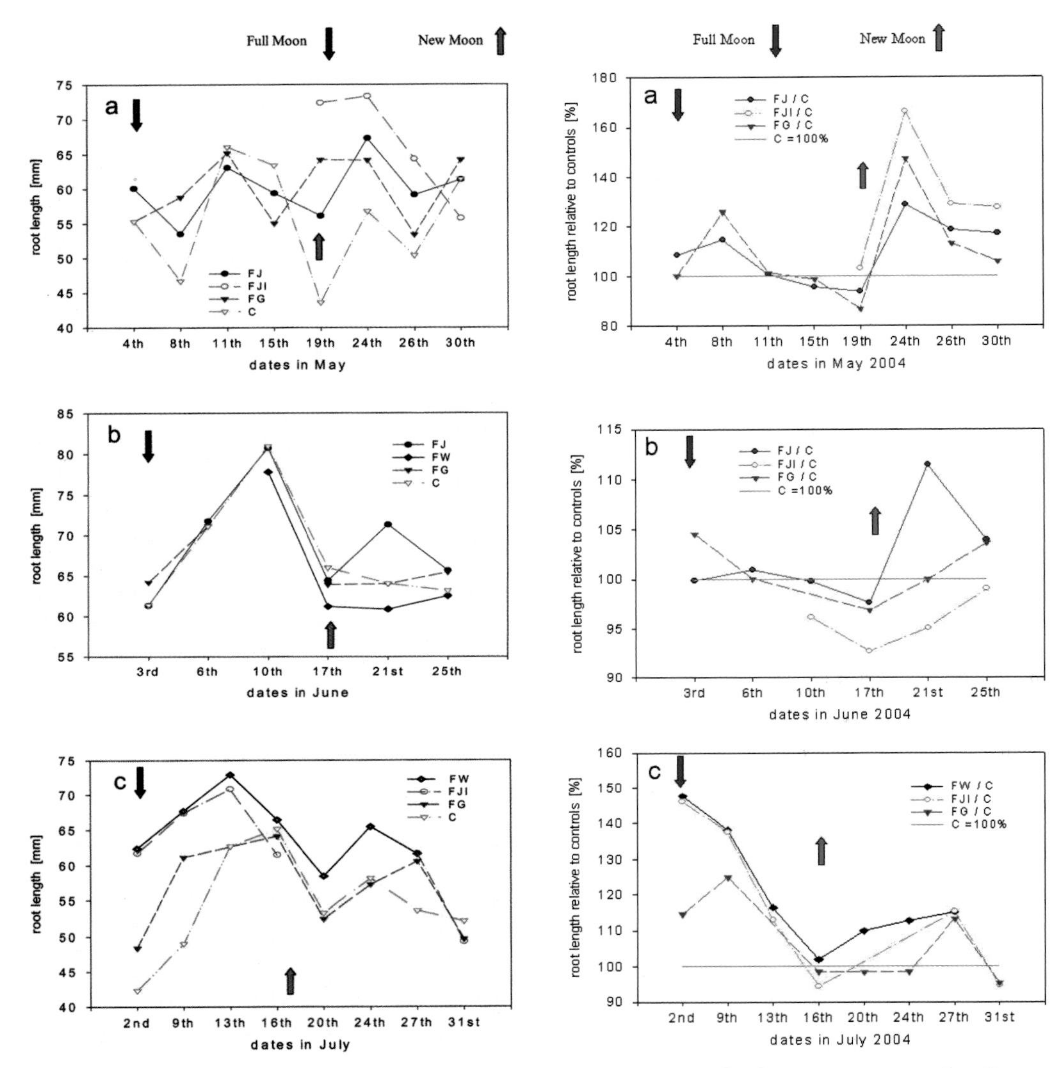

Fig. 45 ***Root lengths*** **through (*a*) May, (*b*) June and (*c*) July with respect to the moon phases**

Fig. 46 ***Length of roots*** **grown in Flowform-treated water divided by root growth of control through (*a*) May, (*b*) June and (*c*) July 2004**

Flowform-treated samples around new moon.

Figure 46 below shows that in most cases Flowform-watered seeds produced the greatest relative increase in growth between new moon and full moon.

This is also the period when root lengths reach a maximum (see Fig. 45), indicating that the Flowform influence might have a tendency to enhance existing growth patterns.

Main results and conclusions

Flowform treatment of water promoted

plant growth, and the difference between Flowform-treated and control samples was more pronounced in root than in shoot lengths.

However, this increase in growth was not constant over the period investigated. The lunar phases were also scrutinized as possible influencing factors causing these fluctuations. The experiments indicated that seedling growth shows a pattern according with moon phases. The relative growth of Flowform samples with respect to the controls reached a minimum around new moon in all three months.

Only by repeating these experiments over a more extended time scale can we draw more precise conclusions.

We hope to take this work further as we continue to use cress growth as an indicator of water quality. Some of this further work might include:

1. Examining the influence of the pump and of the oxygen level on growth. Exploring the dissolved oxygen (DO) levels and establishing controls with identical oxygen levels using a pump and hand stirring.
2. Examining other parameters such as temperature and pH.
3. Experimenting further with the cosmic influences on growth – examining the three aspects of moon position (phases, distance from the horizon and position in the zodiac) and choosing the days of the experiments accordingly.
4. A further investigation could examine the effects of different pumps and different Flowform designs on growth.

Research Project 2: The influence of Flowform treatment on lettuce growth

by Andrea Tranquilini, Heidi Terry and Orit Loyter (2004)

Acknowledgements: We thank John-Wilkes and Arjan Huese for support and advice in this research.

Abstract

Lettuce was grown in the field in the summer of 2004 to investigate the influence of Flowform treatment on irrigation water. The experiment included three lettuce groups where the Flowform Vortex model was used to treat water that was applied: (i) only at the greenhouse stage; (ii) during the whole growth period. Both the above were compared with (iii) controls treated with tap water.

The investigation included a series of observations (of colour, shape, leaf sequence, general appearance, germination rate, leaf size, etc.) conducted once a week. After harvesting the lettuces were observed, tasted, weighed and tested for pH, and differences were detected in most parameters.

The weight of lettuce irrigated with Flowform-treated water increased by 32%, appeared more vibrant in colour and texture and tasted more bitter (the latter related to the type of lettuce chosen). The lettuce that was grown in Flowform-treated water only at the greenhouse stage had characteristics that mostly combined those of the two other groups, though were closer to the lettuce treated throughout with Flowform water.

Introduction

The rhythmic movement of water that can be induced by Flowform technology appears to have a positive effect upon biological processes due to an increase in its sensitivity and oxygen content (Schikorr 1990, Van Mansfeld 1986, Wilkes 2003).

In these experiments, the growth of the lettuce is used as an indicator for the quality of the water as well as to investigate practical advantages of the use of Flowform-treated water as a means of crop irrigation.

We were able to combine different research methods to assess lettuce growth and quality. We performed quantitative measurements as the lettuce was grown and harvested (such as weight, root length, size, pH, soil content).

Materials and methods

The experiments consisted of three groups of lettuce: one grown in Flowform water, the second in tap water and the third grown in Flowform water in the greenhouse, then in tap water after replanting, thus making it possible to observe the influence of the Flowform-treated water at the greenhouse stage and during the whole growth period.

On 6 June, lettuce seeds (type 'Cocarde') were sown in three trays in the greenhouse, each tray containing 120 seeds. The soil used was a mixture of peat and compost formed into soil blocks. The seeds were grown in the glasshouse for a period of three weeks, during which they were watered according to need, usually every other day, with about 0.5 litres of water per tray. All trays were watered at the same time and given the same amount of water. Two trays were irrigated with the Flowform Vortex-treated water and one was irrigated with tap water. On the days of watering, the tub containing the water for the Flowform Vortex cascade was filled with tap water, then the water was circulated through the forms for 60 minutes starting at 7.30 a.m., with a centrifugal amphibious submersible pump P1800 with a flow rate of approximately 80 litres/min. Tap water was used as control.

On 27 June, 21 days after sowing, the lettuce was transplanted consecutively into three plots in the upper garden at Emerson College, Sussex, where the Healing Water Institute is based. Each plot contained 70 lettuce seedlings planted in three rows.

In plot 1, the tap-irrigated seedlings (TT) were planted and continued to be irrigated with tap water. In plots 2 and 3 the Flowform-irrigated seedlings were planted. Plot 2 was irrigated with tap water after planting (FT), and plot 3 with Flowform-treated water (FF). A gap of 1.5 m was kept between plots 2 and 3 to avoid water drainage.

Biodynamic (BD) preparation 500 was sprayed on all crops (made by mixing in Flowform-treated water for all groups). The lettuce was allowed to grow until 23 August. During this time the plots were irrigated when needed (about twice a week). The seedlings were irrigated with the same amount of water (about 35 litres per plot), and irrigation was monitored so that all plots received the same amount of water. There was hardly any rain during the growth period in the field, therefore irrigation accounted for most of the water received by the lettuce.

Throughout the experimental period,

the lettuces were observed once a week. On 23 August, eight weeks after transplanting into pots, the lettuce was harvested and analysed. Soluble contents were measured according to the Brix method.

Results and discussion

At the greenhouse stage, FF and FT received identical treatment; both being irrigated with Flowform-treated tap water (see below). After transplantation into the field, however, their treatment became different in that FF was irrigated with Flowform water and FT with tap water only.

The FF and FT seedlings germinated earlier than the tap water control; the biggest difference (25.6%) was observed 4 days after planting (Fig. 47). However, on day 8 after planting, the mean difference had decreased to 4.3%, until later no differences could be observed.

Roots of three seedlings of the same leaf size were chosen randomly from each of the three trays and measured before transplanting from greenhouse to field, and no differences in root length could be detected.

At this stage, subtle differences were observed when comparing the growing seedlings. The plants treated with Flowform water seemed more delicate in colour and shape, and the control plants were darker and more compact.

Just before transplanting, seedlings in the trays were observed on 27 June as well as once a week while the lettuces were grown in the field. All three trays were filled up completely so that the number of plants was identical. The soil in the trays irrigated for three weeks with Flowform-treated water looked fresher, brown and more alive in comparison with the tap-water tray, which was dry and mouldy (see Fig. 48).

The growing period in the field

The lettuce was grown in the field from 27 June to 23 August. During this period, plants were observed on three occasions (29 July, 2 and 23 August[?]) and the results of the three observations were combined.

Number of leaves (measured July 29): In each plot, 15 plants were chosen randomly, by picking the 6th, 12th, 18th, 24th, and 30th plant in each of the three rows (total of 5 plants per row). For each plant the number of leaves was counted. The average number of leaves was higher for plants treated with Flowform water (8.3 for FF plants, 8.2 for FT, and 7.4 for tap watered TT plants, see Fig. 50).

The mean pH values were 6.8 for the FF plot, 7.1 for FT, 6.8 for TT, and 7.0 in the

	FF Flowform	FT Flowform/Tap	TT Tap
6 June: no. of planted seeds	120 (100%)	120	120
10 June: no. of germinated seeds	53 (44.2%)	55 (45.8%)	43 (35.8%)
14 June: no. of germinated seeds	95 (79.2%)	99 (82.5%)	93 (77.5%)

Fig. 47 *Germination rates* measured on the 4th and 8th day after planting. Total numbers and percentage values in brackets.

Fig. 48 *Soil in Flowform and tap water trays* (27 June)

gap between the plots (with three pH measurements each), indicating that the pH value was not altered significantly in the different plots (Fig. 50).

Phenomenological observation

Phenomenological observations were carried out four times, on 27 and 29 July, and on 2 and 23 August.

The FF and FT (F = Flowform-influenced) plants seemed to be slightly larger on the whole. Most differences though had to do with the way the plants looked. The F lettuce leaves had a lighter, vibrant green colour compared with the darker tap-watered lettuce. The F plants looked younger, livelier and fresher. In the leaf sequence the F lettuce appear

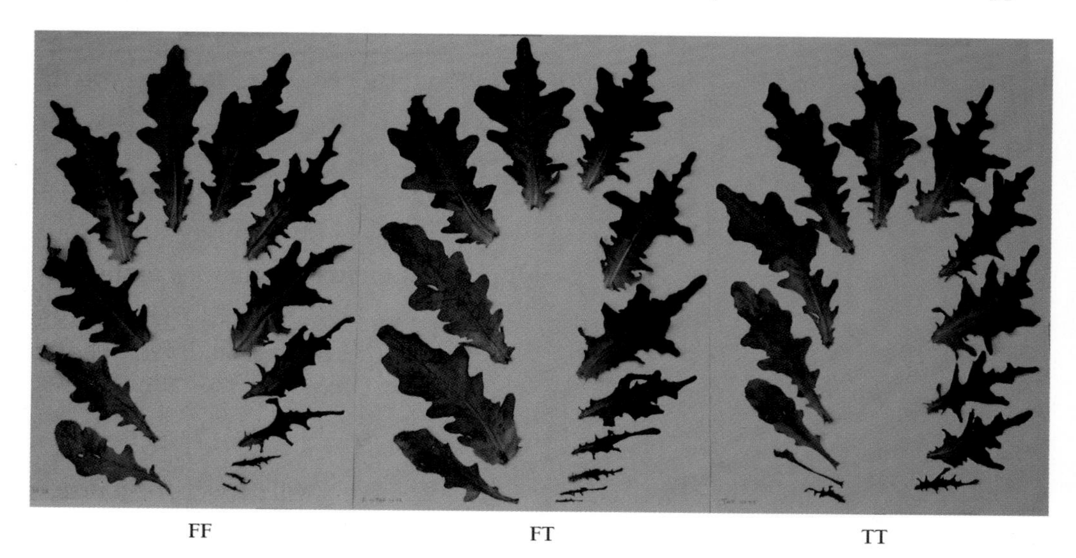

Fig. 49A *Leaf sequences on 4 August*

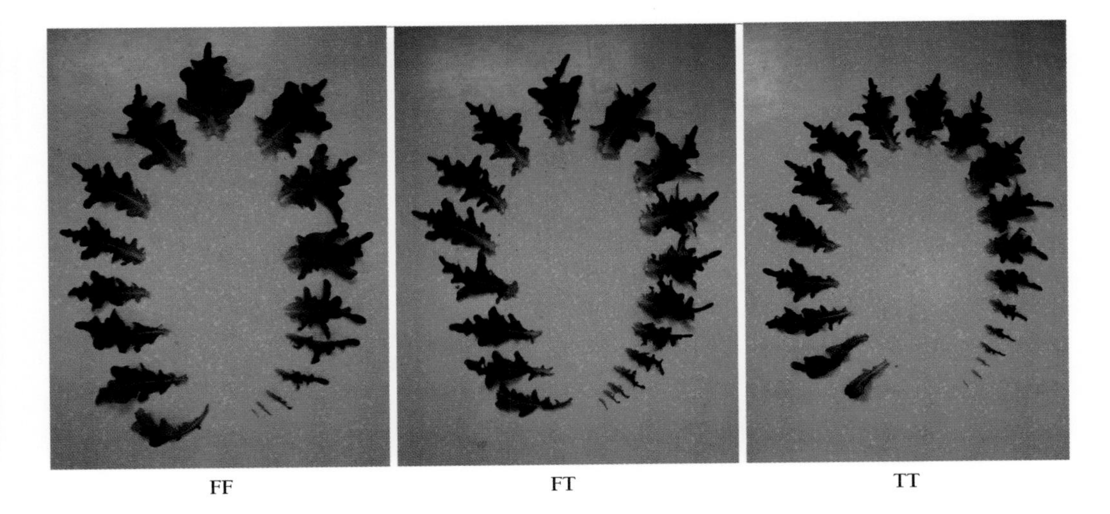

Fig. 49B *Leaf sequence on 15 August*

more expansive and the individual leaves more luscious (three-dimensional), narrower and ordered (Fig. 49A and 49B show leaf sequences). The tap lettuce gave a contractive and more central impression (both Fig. 49A and 49B show the individual leaves). The roots of the F lettuce were longer and had slightly more hairs.

Final analysis of the lettuces after harvesting

	FF Flowform water	**FT Flowform water at greenhouse stage**	**TT tap water**
Number of leaves	8.27 ± 0.54 [SE]	8.20 ± 0.34 [SE]	7.40 ± 0.36 [SE]
Mean pH value of soil	6.8	7.1	6.8
Mean root length [cm]	21.67 ± 0.73 [SE]	18.67 ± 0.88 [SE]	17.83 ± 1.42 [SE]
Mean weight [g]	178.4 ± 8.6 [SE]	168.7 ± 8.0 [SE]	134.7 ± 7.8 [SE]
Total weight [g]	7137	6748	5388
Increase in total weight compared to control TT [%]	32.5%	25.2%	—
Soluble content	1.44 ± 0.19 [SE]	1.68 ± 0.14 [SE]	2.04 ± 0.17 [SE]

Fig. 50 *Data for lettuce plants* irrigated with Flowform water continuously (FF), only at the greenhouse stage (FT), and with tap water (TT). SE: Standard Error. Underlined numbers indicate a statistically significant difference with respect to the TT controls ($P < 0.1$)

General impression

The FF lettuces were larger in size and had more leaves than the FT samples, the TT plants being the smallest of all. No other apparent differences in colour or texture could be detected. Roots appeared to be thicker in FF and FT plants, and the soil seemed to be harder to remove with tap-watered plants.

Lettuce weight

Beginning with the east side of each plot, the first 40 lettuces were harvested from two out of the three rows in each plot. The average lettuce weight increased by 32.5% for FF samples and by 25.2% for FT samples with respect to the plants irrigated with tap water only (Figs 50 and 51).

The distribution of lettuce weight (i.e. the number of lettuces in each weight group) shows that the increase of weight in the Flowform-watered FF and FT groups is due to the fact that these groups have more lettuces of average weight than the TT lettuces (Fig. 52).

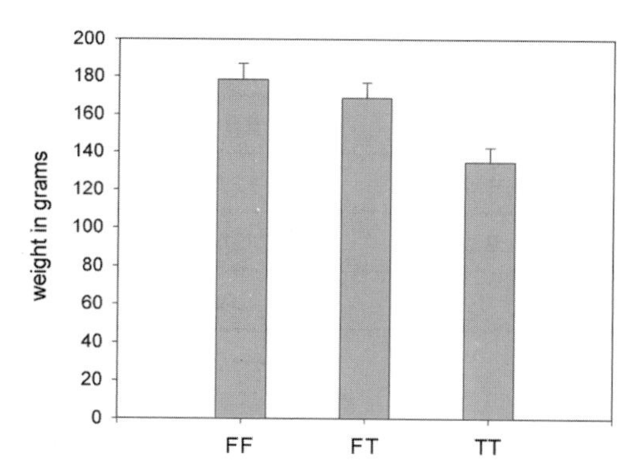

Fig. 51 ***Mean weight of the lettuce samples:*** **FF (only Flowform water), FT (Flowform water, then tap water) and TT (tap water) after harvesting**

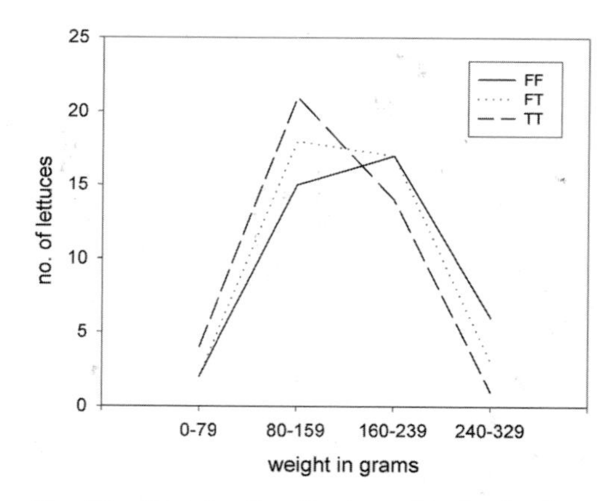

Fig. 52 ***Weight distribution of individual lettuces*****, samples FF, FT and TT**

This is a very desirable quality from the productive point of view. In contrast, most of the tap-watered lettuces (TT) belong to the groups with lower weight.

Soluble content tests

All measurements were conducted on the 7th leaf of 5 lettuces of each type. Soluble contents (measured by the Brix method) were highest in lettuce irrigated with tap water and lowest in lettuce irrigated with Flowform water.

Taste test

Leaves no. 9 and 10 of the lettuces type FF and TT were removed from the plant, washed and placed in two bowls numbered 1 and 2. Fifteen people were asked to taste a whole leaf from each bowl and to wash their mouth between tastes. They were asked to comment on the difference in taste.

Most people commented on the sweetness/bitterness of the leaves. The lettuce irrigated with Flowform-treated water seemed to be more bitter and stronger in flavour and texture than the lettuce irrigated with tap water.

Research Project 3: Influence of Flowform water on wheat growth

Introduction

Indications have been found in several laboratories that water is capable of storing and transmitting information about substances it has been in contact with (Belon et al. 1999, 2004, Davenas et al. 1988, Ludwig 1991).

It has been demonstrated that transmission of chemical information to a solution is possible electronically in the absence of molecule transport (Schiff 1995); and electromagnetic frequencies have been transmitted to water through a sealed ampoule (Gross 2000b).

The spontaneous formation of higher order clusters ('coherent domains') in water leads to structures that vibrate at high frequencies. These structures are referred to as crystalline-fluid, since they have a high degree of order similar to that of a crystal. Coherent domains transmit long-range electromagnetic fields that can lead to qualitatively new effects (Del Giudice and Preparata 1994).

Wolfgang Ludwig (1991) and Peter Gross (2000b) presume that water trans-

Fig. 53 *Rotating table with wheat seedlings*

mits such information to living organisms and, depending on wavelength, these frequencies can be life-supporting or conversely they can be injurious. The information stored in water can be altered or erased either by heating or by repeated rhythmic swirling movements (Gross 2000a).

Flowform cascades generate such rhythmic lemniscatory and vortical movements. This research project aimed to investigate the effect of Flowform-treated water on plant growth. The influence of several geometrical 'sheaths' around a Flowform cascade was evaluated.

Materials and methods

To investigate the effect of Flowform water on plant growth, wheat seedlings were grown in different water samples treated with Flowform cascades. All samples used the same equipment and the same brand of centrifugal pump (Kockney Koi Yamitsu submersible fountain pump, model KKYFP2400) for one hour's water circulation at a flow rate of 6 l/min. Only for the Flowform Vortex model, which requires a much higher flow rate (200 l/min, FV in Fig. 15), was an in-line Hidrostal helical screw pump used.

Biodynamic wheat seeds from a harvest in 2004 from Tablehurst Farm, Sussex, were used for the experiments. Twelve wheat seeds were placed equidistantly on a mesh that was fixed inside each glass containing the water sample (Figs 53 above and 54 below).

All glasses were placed on a rotating table to ensure even light distribution. Wooden divisions and a central plastic pipe separated different samples so that they were not contaminated by each other's presence.

The roots grow through a nylon mesh into the water. After 8–16 days, the seedlings were photographed and analysed. On the day of the experiment, water samples were used for capillary dynamolysis.

Fig. 54 *Wheat seedlings germinating* on top of nylon mesh after 2 days (left) and 4 days (right)

Measurement and statistical analysis

After a growth period of 8–16 days, the results were photographed and root lengths were measured from digital images using NIH-image software and the macro described in Research Project 1 (Fig. 55).

In all experiments, T-tests were carried out, and results were considered significant if P was less than 0.05. Mean growth was calculated using only the germinated seeds.

Results and discussion

Set 1: Flowform cascade inside a spherical icosidodecahedron

In two experiments (9 and 10 August 2006), a laboratory stackable cascade consisting of 16 units was hung inside a copper tube model of a spherical icosidodecahedron (FID) about 2 metres in diameter, to compare effects with a corresponding Flowform cascade (FG, Fig. 56).

Control treatment (C) was carried out simultaneously by circulating the tap-water sample through the same make of electric pump to equalize oxygenation.

The model was constructed and loaned by Tibor Adoryan. We were only able to use it for two days, so we had no control over the timing of the experiments.

Generally, root growth of seedlings was promoted in Flowform-treated (FG) water compared to the control samples (Figs 57–58).

In one experiment (9 August), the projected polyhedron (FID) further increased the growth-enhancing effect of the Flowform (FG) cascade (Figs 57, 58 and 59).

Fig. 55 *Glasses containing wheat seedlings growing in water samples* from different treatments, showing overlaid black size-markers using NIH-image software to demonstrate the measurement technique.

Fig. 56 Laboratory stackable cascade (FG, left), and copper model of the spherical icosidodecahedron (FID, right)

Fig. 57 Wheat seedlings (experiment 9 August) after being grown for 9 days in control water (C), water from the stackable Flowform cascade (FG) and from a similar cascade placed inside the polyhedron (FID)

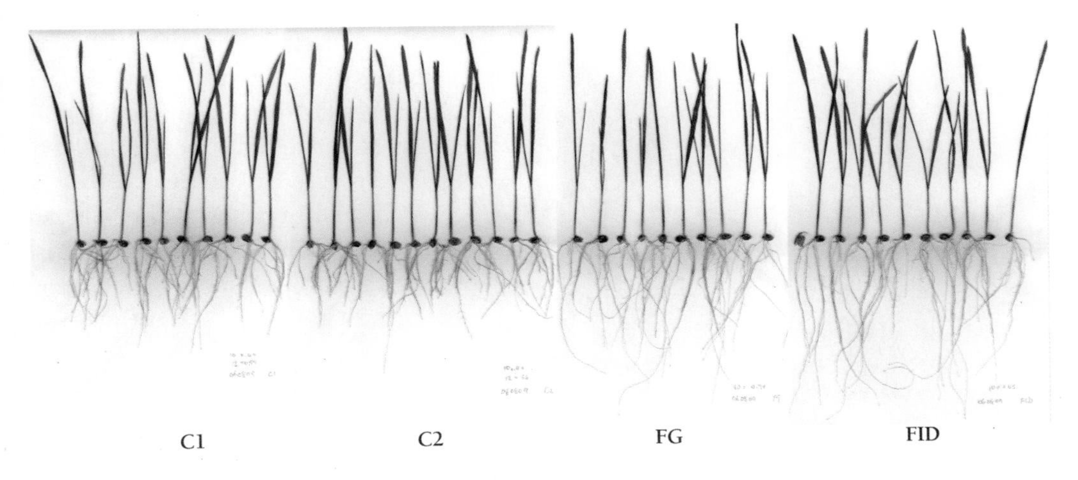

Fig. 58 Pressed wheat seedlings after 9 days showing enhancement of root growth in Flowform samples FG and FID compared to controls C (experiment on 9 August)

In contrast, in another experiment (10 August) the cascade with the polyhedron (FID) resulted in reduced root growth compared to the Flowform cascade FG (Fig. 60).

In total (experiments on 9 and 10 August), water treated with cascade FID resulted in an overall 23.1% increase of mean root length compared to the controls, and the largest difference could be detected in plants treated with water from cascade FG, resulting in an increase in root length of 35.3% (Fig. 60).

Root lengths of the Flowform samples FG and FID are statistically different from the controls at the 0.1% level ($p<0.001$).*

Generally, shoot lengths did not display any detectable difference.

Images from capillary analysis show a more vigorous pattern in FG and FID samples compared to controls C, and the middle zone is extended (Fig. 61).

Thus, in all examples, root growth was consistently promoted in FG samples compared to controls, whereas the polyhedron had an inhibiting effect in one (10 August), and an enhancing effect on root growth in another experiment (9 August) with respect to FG samples (Figs 44 to 46). Notably, 9 August 2006 was full moon and, according to the *Biodynamic Sowing and Planting Calendar*, a 'root day' as well (Maria and Matthias Thun 2006).

---

* **Statistical significance:** A result is called 'statistically significant' if it is unlikely to have occurred by chance. The level of significance á (alpha) gives an indication of the probability that a result could have occurred merely by chance. Popular levels of significance are 5%, 1% and 0.1%. If the P-value is less than the significance level, then the result is considered statistically different. For example, if there is only one chance in a hundred that a result could have happened by coincidence, a 1% level of statistical significance ($P = 0.01$) is implied. Typically it is asserted that the results are 'statistically significant' if a significance test gives a P-value lower than 5% ($P < 0.05$). The smaller the P-value, the more significant the result is considered to be.

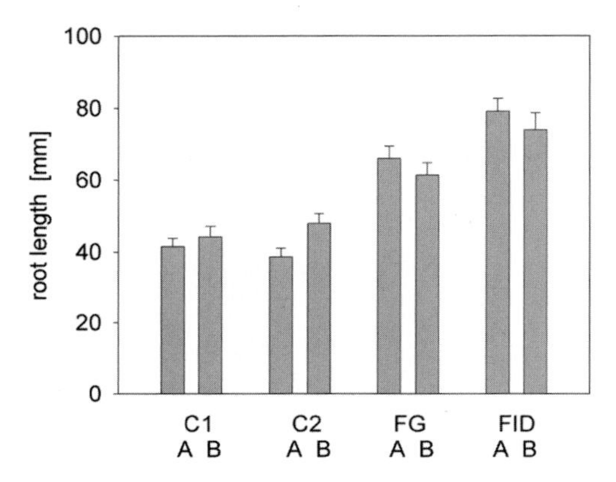

Fig. 59 Mean root length in mm (experiment 9 August 2006), *C1* (n = 60), *C2* (n = 60) controls, *FG* Laboratory stackable Flowform cascade (n = 60), *FID* second cascade inside a copper tube model of a spherical icosidodecahedron (n = 61), *n* number of roots for each treatment, *A*, *B* data for roots inside individual glasses, bars with standard errors

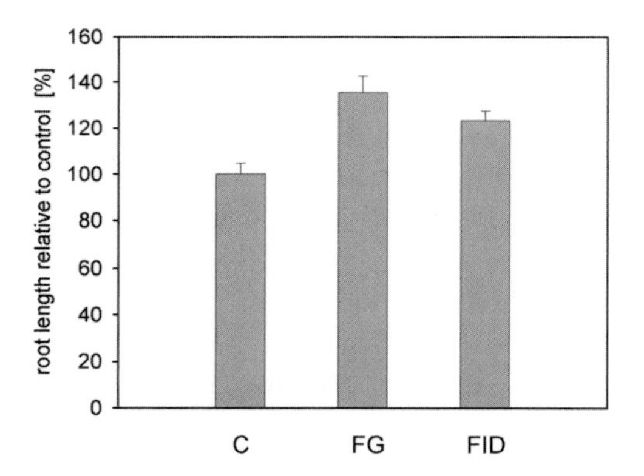

Fig. 60 Experiment on 10 August 2006. The conditions had changed. It was no longer a root day and was also considered unfavourable for sowing (Thun *Calendar*), with moon in perigee moving from Waterman into Fishes. Due to the rhythmic sensitizing of the water these influences are readily taken up by plants; the results tended to be more irregular, and experimental factors may also have had an influence

Given an opportunity in future to carry out extended experiments over a longer period, the results could be more interesting. Because conditions are constantly changing in nature we can never expect a merely mechanical repetition of results. In particular, the results above from 9 August show interesting potential for such treatments (Figs 59, 60).

Set 2: Rocker Unit and Flowform cascades inside projected wooden and bronze icosahedra

In several experiments (2, 9, 15, 23 May, 25 September and 1 October 2007), seedlings were grown in water circulated for one hour through a cascade of 16 Flowform laboratory stackable models (FG), and through similar cascades placed partly within a projected icosahedron made of bronze rods (FGIM) and hanging completely inside a wooden projected icosahedron (FGIW, Fig. 62). In two of these experiments (25 September and 1 October), in addition, water was moved rhythmically in a Rocker Unit (Fig. 66 right), without using an electrical pump. Control samples (C) were grown in tap water circulated for one hour with a submersible centrifugal pump.

In this set of experiments, extending over 5 months, results variability was relatively high.

In some cases (experiments of 2 and 9 May), water from control samples (C) and from the cascade inside the metal icosahedron (FGIM) resulted in much shorter roots which scarcely grew into the water, whereas roots from all other Flowform samples were long, dense and abundant (Fig. 63).

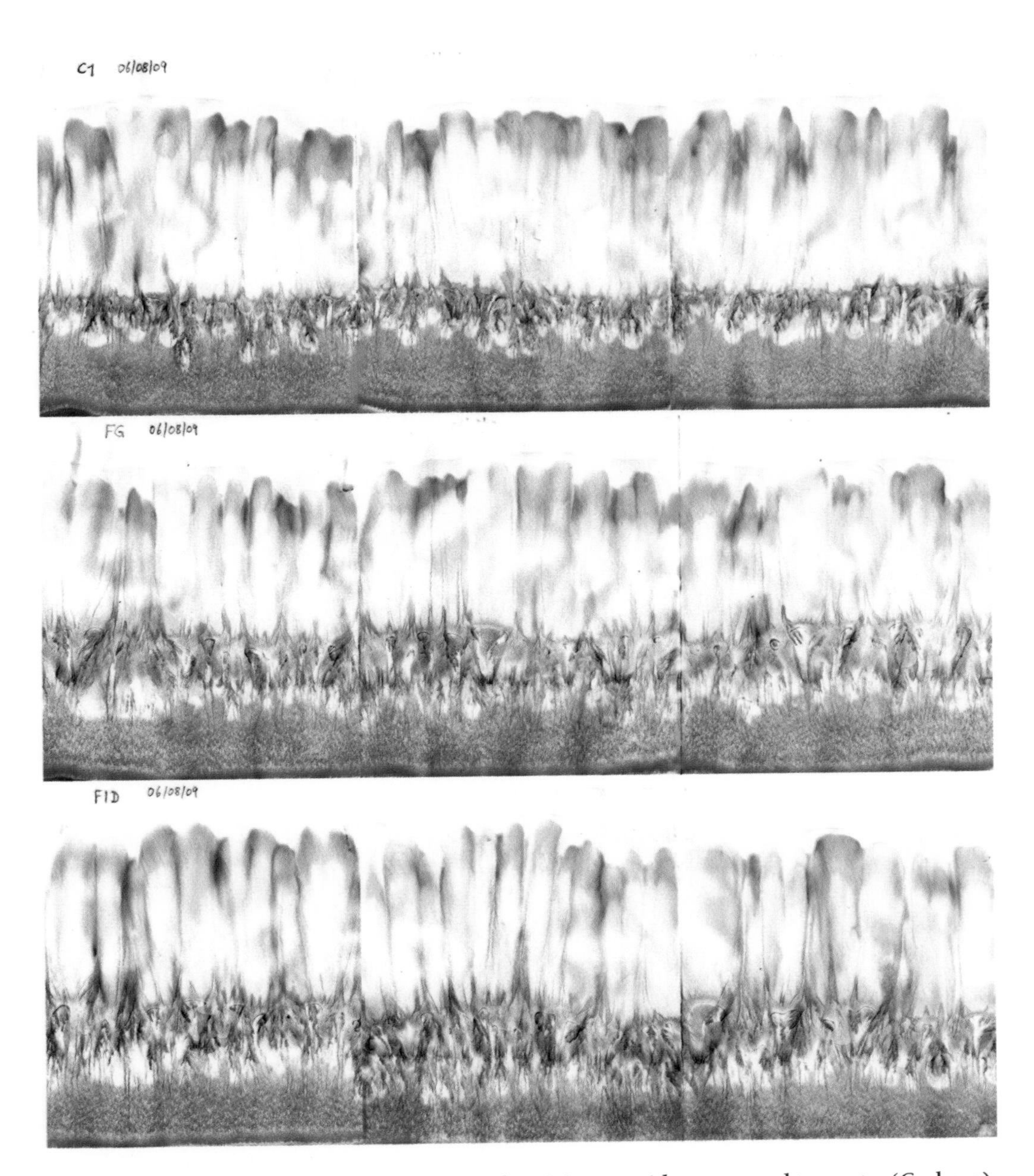

Fig. 61 Capillary images from samples treated on 9 August with oxygenated tap water (C, above), with a laboratory stackable cascade (FG, middle) and with a corresponding stackable Flowform cascade hanging within the icosidodecahedron (FID, below). All images show a more vigorous pattern in Flowform samples FG and FID compared to the controls C.

Fig. 62 *Flowform cascades from left to right:* laboratory stackable Flowform cascade (FG); inside a projected wooden icosahedron (FGIW); partially inside a projected metal icosahedron (FGIM)

Fig. 63 ***Wheat seedlings 20 days after exposure to control water*** **(C), to water from a Flowform stackable cascade (FG), and from a similar cascade placed inside a metal (FGIM), and a wooden projected icosahedron (FGIW). (Experiment of 9 May)**

In other experiments the differences were much more subtle, as shown in Fig. 64 (experiment of 25 September). In one experiment (23 May) the result was reversed, in that controls resulted in the longest roots compared to all Flowform treatments.

This raises the distinct idea that in botanical processes subtle influences are at play, and that sensitized water can tend to emphasize such influences.

Experiments over a longer period show that rhythmic processes generated through Flowform activity have a sensitizing effect on water.

In many of the tests carried out, the water may be more readily influenced by the growth-inhibiting or growth-enhancing environmental and cosmic influences at that time.

Again, this indicates that we need to become much more aware of when we carry out planting and harvesting activities.

Fig. 65 below shows the mean values of all experiments.

Root growth was enhanced by 38.8% in the seedlings treated with water from the cascade within the wooden icosahedron (FGIW), by 20.1% from the laboratory stackable cascade (FG) and by 25.3% from the Rocker Unit (FR).

However, root growth of FGIM-treated plants (bronze icosahedron) was inhibited by 22.7% with respect to the controls.

All samples are statistically different from controls at the 1% ($P < 0.01$) level (Fig. 65 below). Generally, shoot lengths did not show any detectable difference.

Fig. 64 ***Wheat seedlings 13 days after exposure to control water*** **(C), to water treated with a Flowform stackable cascade (FG), and from a similar cascade placed inside a metal (FGIM), and a wooden projected icosahedron (FGIW). FR treated with Rocker Unit. (Experiment of 25 September)**

Set 3: Flowform stackable cascade inside a bronze icosahedron, and a cascade of 38 Flowforms

In several experiments (6 and 7 January 2005) the following samples were compared:

1. **Control** C was tap water moved through a pump for one hour and allowed to fall into itself to create turbulence and oxygenate. Samples were collected from the pump outlet.
2. **FW** cascade consisting of 38 units of the laboratory stackable model through which water from the mains hose was allowed to flow once only (see Fig. 42 right). Samples were collected directly from the cascade into glasses.

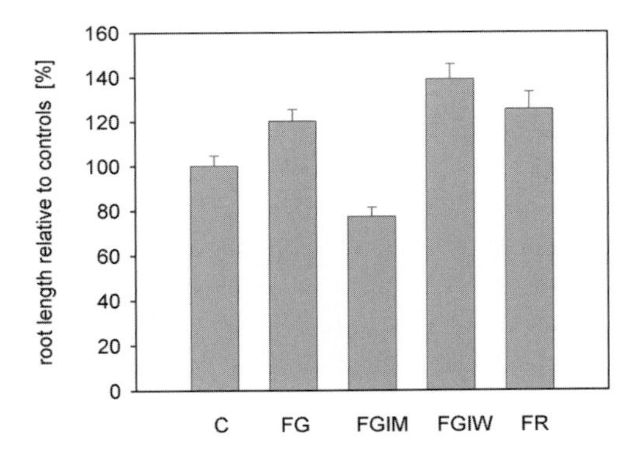

Fig. 65 ***Mean root length in % with respect to control treatments*** **(*C* = 100%, n = 284), *FG* cascades of 16 laboratory stackable models (n = 260), *FGIW* similar cascade of 16 forms inside a projected wooden icosahedron (n = 279), and *FGIM* partially inside a projected metal icosahedron (n = 219), *FR* rocker Flowform (n = 84), bars with standard errors, n number of roots per sample**

3. **FJI** cascade consisting of 16 laboratory stackable models hung partly within a projected icosahedron made of bronze rods. The elliptical projection is 100 cm high and the cascade 150 cm high. Water was circulated for an hour by centrifugal pump at 6 l/min. Samples were taken directly from the cascade into glasses (see Fig. 42, centre).
4. **FV** cascade consisting of five Flowform Vortex vessels above a 2000 litre tank (Fig. 66 left). Water was circulated for one hour at 200 l/min using an in-line Hidrostal helical screw pump. Samples were taken directly into glasses from the cascade outlet.

Generally, root lengths of plants in different glasses of the same treatment sample were comparatively similar (Fig. 67).

In this set of experiments, Flowform treatment resulted in a dramatic increase in root lengths. Root lengths of plants grown in water treated with the Flowform Vortex (FV) increased by 307% compared to the controls. The cascade within a projected icosahedron (FJI) led to an increase of 112%, and the cascade (FW) promoted root growth by 48% (Fig. 68).

However, since the water used for the Flowform Vortex cascade was taken from the tap earlier than for the other samples,

Fig. 66 The Flowform Vortex (FV, left) and the Flowform Rocker (FR, right)

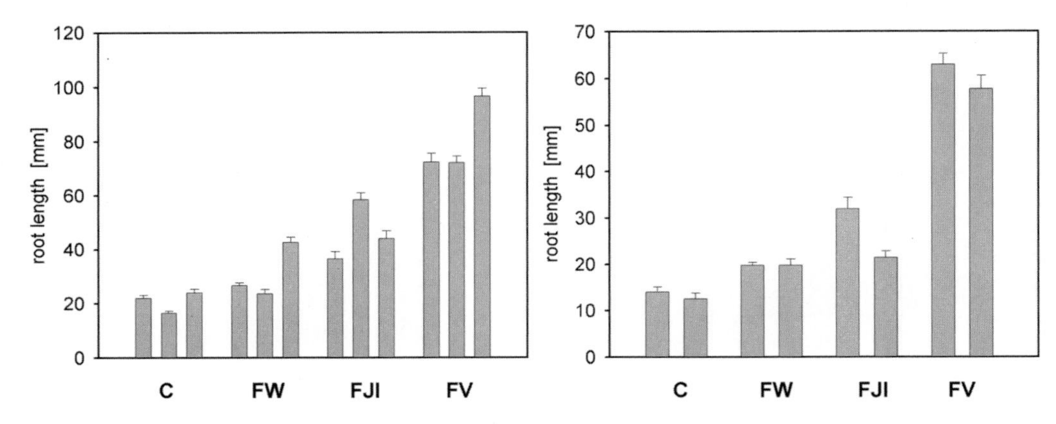

Fig. 67 *Mean root length in mm:* left, Experiment 7 January, three replicas (glasses). Right, Experiment 6 January, two replicas for each sample, each bar represents mean root length in a single glass, *C* control, *FW* cascade of 38 laboratory stackable units, *FJI* cascade of 16 laboratory stackable units partially inside a projected metal icosahedron, *FV* the Flowform Vortex, bars with standard errors

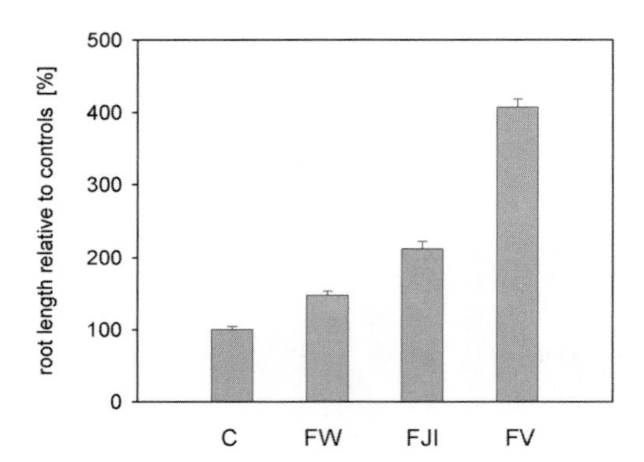

Fig. 68 *Mean root length in % with respect to control treatments* (*C* = 100%, n = 76), *FW* cascade of 38 Flowform laboratory stackable units (n = 121), *FJI* cascade of 16 Flowform laboratory stackable units partially inside a projected metal icosahedron (n = 127), *FV* the Flowform Vortex cascade (n = 139), bars with standard errors, *n* number of roots in each sample

and stored in a large open container, the results of the Vortex cascade have to be considered with caution and need to be repeated to assure conclusive results.

All samples in Fig. 68 are statistically different from controls at the 0.1% ($P < 0.001$) level.

To summarize, in most experiments with wheat seedlings, root but not shoot growth was promoted when grown in Flowform water. An exception was the water from the metal (bronze) icosahedron, which inhibited root growth.

Promotion of growth was always statistically significant at the 5% level (in some cases also at the 1% or 0.1% level).

Generally, the variability of results in single experiments was relatively high, which could be an expression of seasonal and/or cosmic influences.

All the experiments need to be conducted over extended periods and could be related to planetary phases to enable us to ascertain whether certain periods of enhancement or inhibition of root growth come into play.

Research in the near future

The Healing Water Institute has formed connections with more than 30 countries, and is established as coordinated legal entities in England, New Zealand and the USA. At present there are active connections with 20 countries relating to a wide variety of practical, scientific and artistic projects.

Funding for research and vital administrative planning and coordination is being sought for the institute in England as well as that in New Zealand. The Healing Water Institute in the USA is focusing primarily on educational aims, but is able to fund-raise on behalf of these institutions.

Further projects are planned in relation to the following areas of research and practical application:

- We need *to establish more advanced test systems* to further examine the effects of Flowform technology on water quality. The institute has already conducted experiments with plants that can easily be grown under controlled conditions (such as cress, wheat, etc.), enabling us to examine the effects of Flowform water on growth and vigour.
- We need *to further analyse and repeat our plant experiments* with cress (Exp. 1) and wheat (Exp. 3) seedlings, as well as lettuce plants (Exp. 2). These have already shown that there is a significant effect of Flowform-treated water on the vigour and growth rates of different plants.
- We need *to investigate the influence of lunar and planetary phases* in relation to different Flowform cascades and geometric envelopes. There is much data available and this needs analysis according to experimental dates. Because organic phenomena operate in living rhythms (unlike inorganic matter) we need to see what rhythms are influencing our experiments.
- We will *use the capillary method, crystallization and drop-picture* techniques, in connection with quantitative analysis, to determine the effects of different rhythmic treatments and different Flowform design types.
- The *drop-picture method is currently set up at the institute* to be used as a complementary qualitative method for the analysis of water quality.
- *Capillary and crystallization tests are already routinely used* at the institute to characterize water in relation to its life-supporting influence and to investigate the effects of Flowform treatments. We want to carry this work much further and learn how to read and interpret in more detail the crystallization and capillary images, so that reliable statements can be made about the qualities of the different water samples and their effect upon plant growth.
- We want *to examine how polyhedra of different designs and materials might influence,* and have an enhancing effect upon the Flowform sensitizing rhythmic function. We have started conducting experiments testing plant growth with Flowform water from cas-

cades inside several projected polyhedra (see Fig. 42 and Fig. 62).

- It is important to extend our research by examining *micro-organisms' biological responses* to Flowform water, especially regarding effluent treatment.
- We also *intend to further develop the Virbela screw pump* in relation to water lifting applications, for example with biodynamic preparations. The leading concept of this current project is to transport water upwards in an Archimedean manner via an open spiralling channel in such a way that it passes over path-curve surfaces in an expanding and contracting path through left- and right-handed vortices. This lifting technique facilitates a treatment process in a small space without the negative influences of conventional pumping or electromagnetic polluting frequencies.
- *We will investigate the effects of the Rocker Unit* design on water. This is a single form consisting of four cavities in a symmetrical arrangement, capable of applying a rhythmic treatment on a smaller scale without using electrical pumps.
- *The pioneering geometrical path-curve research* of George Adams and Lawrence Edwards needs to be carried further. We want to investigate moving water over path-curve surfaces by further incorporating such surfaces into Flowform vessels.
- *Flowform rhythmic treatment of various agricultural fertilizers, including compost teas and the catalytic biodynamic preparations* needs more investigation both in the laboratory and in the field, to extend existing positive results.
- *Flowform influences in aquaculture and horticulture will* be studied in practical situations, to posit more precise hypotheses for future experimentation.
- *Flowform designs ideally suit fish passes,** and prototyping work in New Zealand needs to be continued, and implemented in working installations.

At a time when water is struggling to sustain the enlivening role it has played for aeons, the work of the Healing Water Institute is unique. It not only researches the life-supporting, energetic capacities of water, amongst other water issues, but also further develops Flowform and other new eco-technologies to practically help water support life.

In addition to these design research tasks, the Healing Water Institutes also engage in educational activities to nurture understanding of water's creative secrets, thus promoting and stimulating essential action and help.

Our intention is to develop the activities and influence of the Healing Water Institutes internationally, from our bases in England, New Zealand and the USA, with the sole aim of helping water, and through it human communities and nature.

* Fish passes are stepped water passes ('fish ladders') to enable fish otherwise blocked by dams and other built structures to return upstream.

Appendix 1
THE HEALING WATER INSTITUTE: HISTORICAL BEGINNINGS

We must learn to see in Nature not only what is already in existence (and therefore dying) but also what is newly becoming in her life. We have to liberate imagination from the bondage of the finished forms of space.

George Adams (1894–1963)

The institute referred to throughout this book was originally founded in 1975 by John Wilkes, Nick Thomas and Nigel Wells. At the beginning of 2010 we received confirmation of our official charitable status, registered number 1133741, under the name Healing Water Foundation (see p. 38, footnote). We can now receive direct funding towards our research in the UK, New Zealand and the USA. We have a growing number of other colleagues around the world interested in rhythm research based on water-quality treatment functions of Flowform technology.

In recent years the work has come full circle. Some aspects of the initial and underlying geometrical themes have repeatedly surfaced in our international seminars. Surprisingly, however, the polyhedra discussed below have also now become directly related to certain applications of Flowform technology, namely their use as protective envelopes within which vertical cascades can be hung. Some initial investigations and results are described in these pages.

There are two main geometrical themes that have been integrated into the Healing Water Institute's original rhythm research: polyhedra and path-curve geometry. Having worked with George Adams to produce his path-curve research apparatus John Wilkes was able to integrate this with the Flowform method in 1970, thus enabling the development of his research.

Rather than examining this in full detail, we wish only to draw attention to its existence here. Anything beyond this exceeds the remit of the present book. Our aim is solely to help readers appreciate, at least, the beauty and value of the applications. A more detailed publication may be possible at a later date.

Polyhedral projections
by John Wilkes

I wish to bring this compilation on Flowform water research to a close by describing some aspects of very early studies in which I was involved under the tutelage of George Adams. My move to London in 1953 to attend the Royal College of Art Sculpture School brought me increasingly into contact with the work of Rudolf Steiner through people who had known and studied under him. I was in the process of writing my final thesis and came to George Adams for advice. My theme was 'The Living Nature of Form'. Among many other subjects, we were studying classical geometry with architect Sergei Kadleigh; and when I mentioned this, George Adams generously offered to instruct me in some aspects of modern synthetic (in contrast to analytic) projective geometry. This geometry is not concerned with number but with process, and is ideal for applying to the study of nature.

We began by studying polyhedra and methods of bringing them into movement through, for instance, the conic sections, sphere, ellipsoid, paraboloid and hyperboloid.

For this purpose one needs to determine the origin of such three-dimensional forms within the two-dimensional plane. All form archetypes are two-dimensional.

For instance, a cube (hexahedron) is generated from three points at infinity (within the Harmonic Net). If we look at it edgewise we see three sets of four parallel lines, indicating the three directions of space. Notice the corners of the cube.

Very soon we find that not only all the other platonics sit within the cube but a multitude of increasingly complex polyhedra are potentially present even if not physically manifest (Fig. 69). They all originate in the same plane at infinity, creating an increasingly intensive 'field of form' as one approaches the infinity within; we could also describe this as the central point of the whole system. Everything physical occurs between the periphery which expresses total expansion, and a central point which expresses total contraction.

If we now imagine the flat plane at infinity moving inwards, this will take place from the physical direction we chose; the spherical envelope of the cube moves out in the opposite direction, through the conic

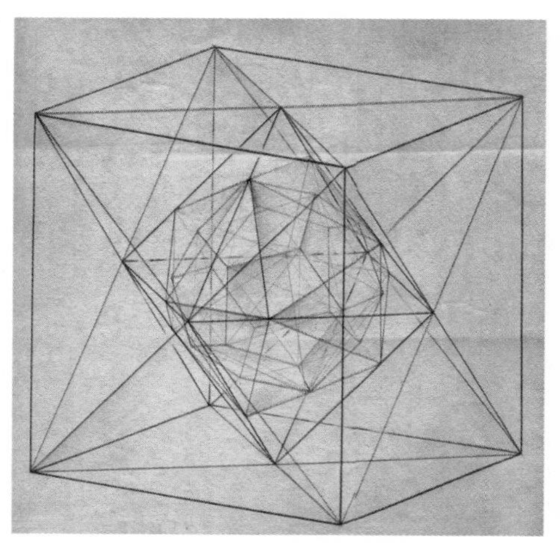

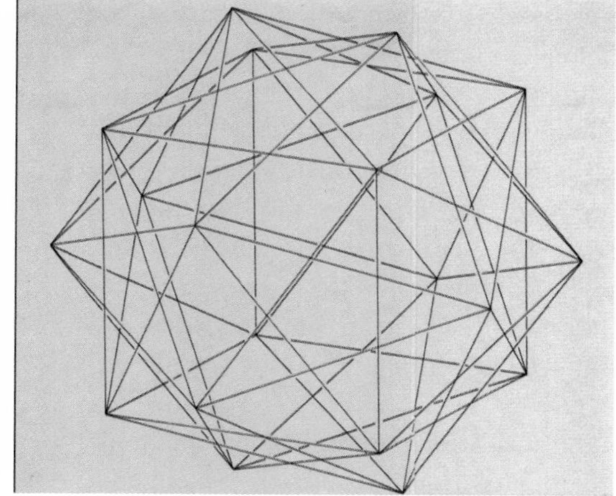

Fig. 69 *Left:* All platonics (drawn) and a multitude of other polyhedra sitting within a cube. *Right:* Pentagram dodecahedron as five cubes

sections, while the infinite point within remains fixed. First of all the sphere becomes slightly ellipsoidal. As the ellipsoid extends outwards with increasing speed its extremity eventually reaches infinity and becomes a paraboloid, beyond which it becomes a hyperboloid. This circumscribing envelope carries within it the polyhedral projections. A fascinating aspect of this process is that we have one spherical and one parabolic form, while ellipsoid and hyperboloid move through a vast range of forms (Fig. 70).

Described from a planar point of view we see that the circle only occurs at right angles to the cone axis: as this angle changes from '0' the ellipse shoots out, becoming longer until the angle is parallel to the cone, when the parabola appears. Beyond this, the plane of the hyperbola cuts both extremities of the cone, moving through its mid-position when parallel to the cone axis.

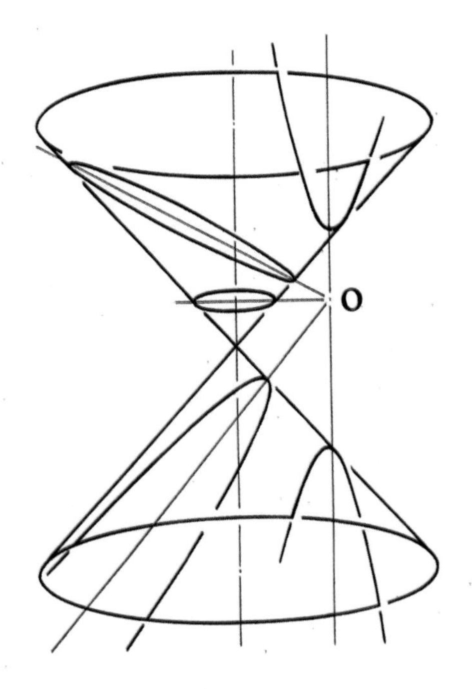

Fig. 70 Sections of the cone

On the left, in Fig. 71, a projection process is shown from a fixed set of three points, which will grow through the conic sections. The beginning is similar to the opening of a bud.

If the projection is rotated towards the left (anticlockwise), while keeping the plane of origin in place, the point 'a' on the right will move out towards infinity. When this is reached the nature of the projection will be as in the central drawing 'a ∞' (Fig. 71). Likewise, if this projection is rotated from front to back (anticlockwise), the nearest point 'b' at the front will move out to infinity 'b ∞' (Fig. 71 right). When this is reached the third drawing will appear with one physical point of origin and two no longer physical but infinitely far away 'a ∞' and 'b ∞'. If the plane in which this last point is resting is now moved out to infinity, the projection can become a regular cube as in (Fig. 69).

The left-hand drawing in Fig. 71 shows the possibility of growth while the origin remains fixed. By means of rotation the plane of origin changes, with the projections taking on a metamorphic relationship. The reason for this is that a change takes place within the two-dimensional origin, which represents the non-physical plane at infinity. When re-entering the three-dimensional, the projection has changed.

Let us look for a moment at the plant, a different realm altogether.

If we consider one leaf, growth which is essentially continuous brings about changes in the shape of the physical sub-

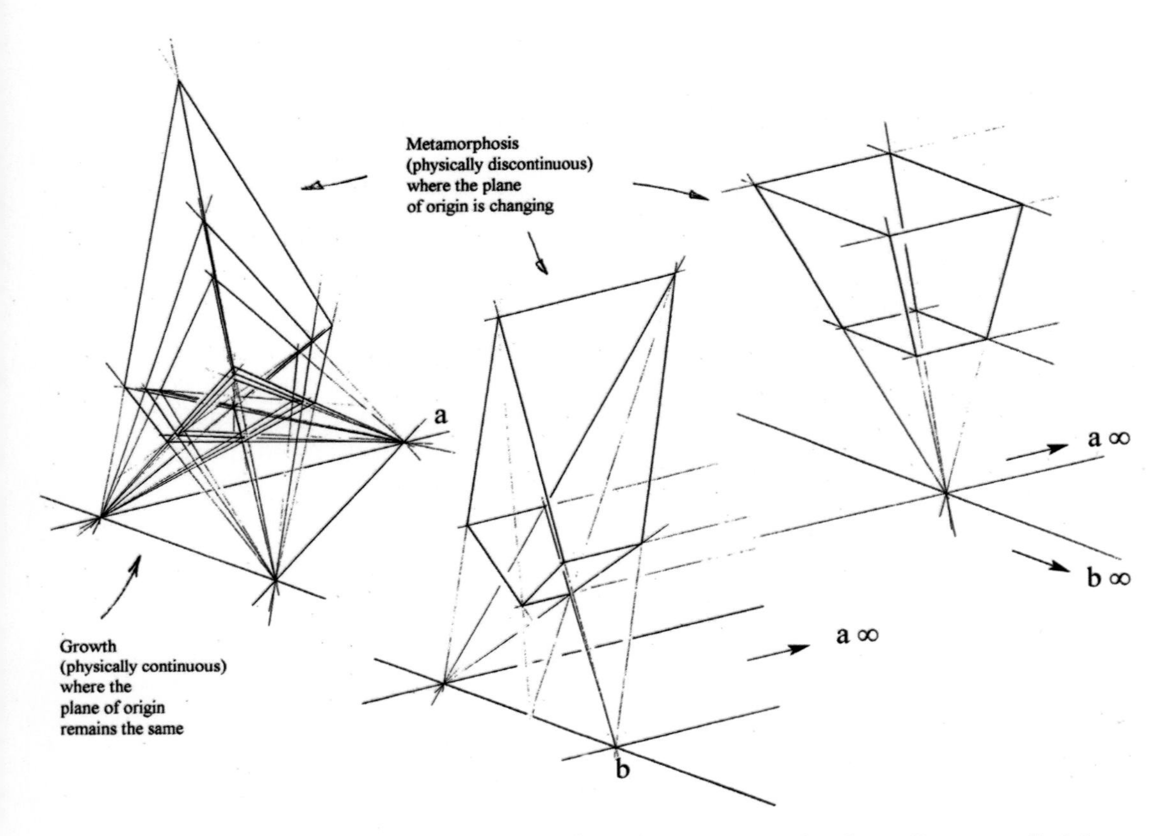

Fig. 71 ***Left:*** **Projection process starts from three points in the plane of origin (called the Harmonic Net or 13 Configuration) and shows stages of growth, while the three points remain fixed. Rotation of the projection anticlockwise leads point 'a' outwards towards infinity as in the** ***middle*** **drawing (a ∞); rotation of this projection from front to back anticlockwise will move point 'b' out to infinity,** ***right*** **drawing (b ∞)**

stance (Fig. 72b bottom). If we look at the next leaf, something much more dramatic has taken place in the non-physical gap, which can bring about a new leaf that also grows but shows quite different characteristics. Growth can be described as a physically continuous process whereas metamorphosis is essentially a physically discontinuous process (Fig. 72b top).

This freehand blackboard drawing (Fig. 73) shows further possibilities with projections. Here the three main symmetries are illustrated with point, line and plane plus, on the right, the regular-type cube projection from the three infinite points (this last drawing is not quite correct in the proportions but suffices to show the principle).

The following figures demonstrate the Harmonic Net or 13 Configuration which provides the origin for the cube (hexahedron), octahedron and rhombic

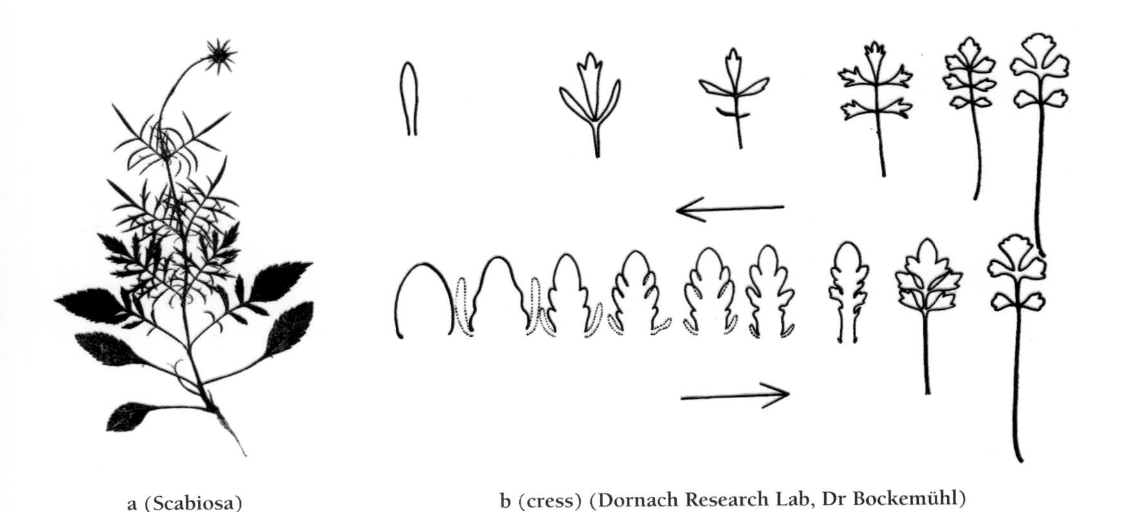

a (Scabiosa) b (cress) (Dornach Research Lab, Dr Bockemühl)

Fig. 72 *(a) The whole plant*, showing the metamorphic development of leaves. *(b) Upper series*, illustrating the process of metamorphosis from right to left where we imagine the archetype is changing. Here we experience a fixed number of stages, visible in space. *Lower series*, continuous growth of the cress leaf, finally reaching the stage of the adult leaf (right). One can imagine the archetype remains the same. Any number of drawings can be used to express this growth process in time. It is hidden from our consciousness; we are only aware of the final result. Early stages (left) are greatly enlarged to allow them to be visible.

Fig. 73 Blackboard drawing illustrating the main symmetries with point (left), line (second from left) and plane (third from left), and the projection of a regular cube (right).

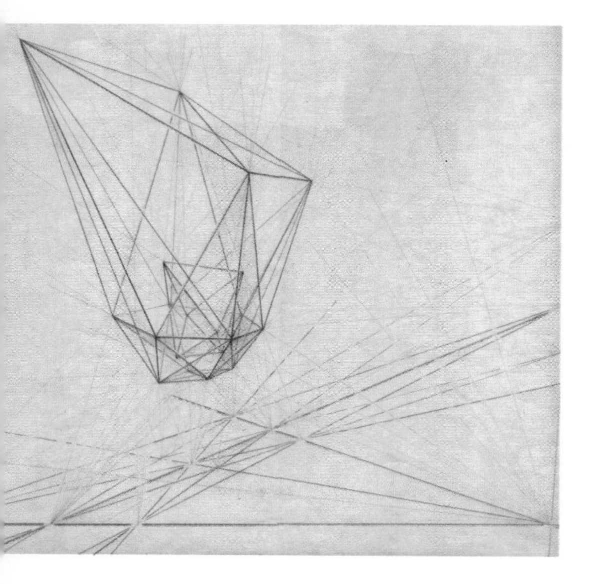

Fig. 74 *Left:* Harmonic net with projected cube in green, octahedron in blue and rhombic dodecahedron in purple
Below: Ink drawing with differentiated lines to distinguish between the three lines, four lines, and six lines. A cluster of rhombic dodecahedra is constructed from the four points, which relate to the six lines

With some study and practice the results are well worth the effort. (Many books on projective geometry, at varying levels, are available for further information.)

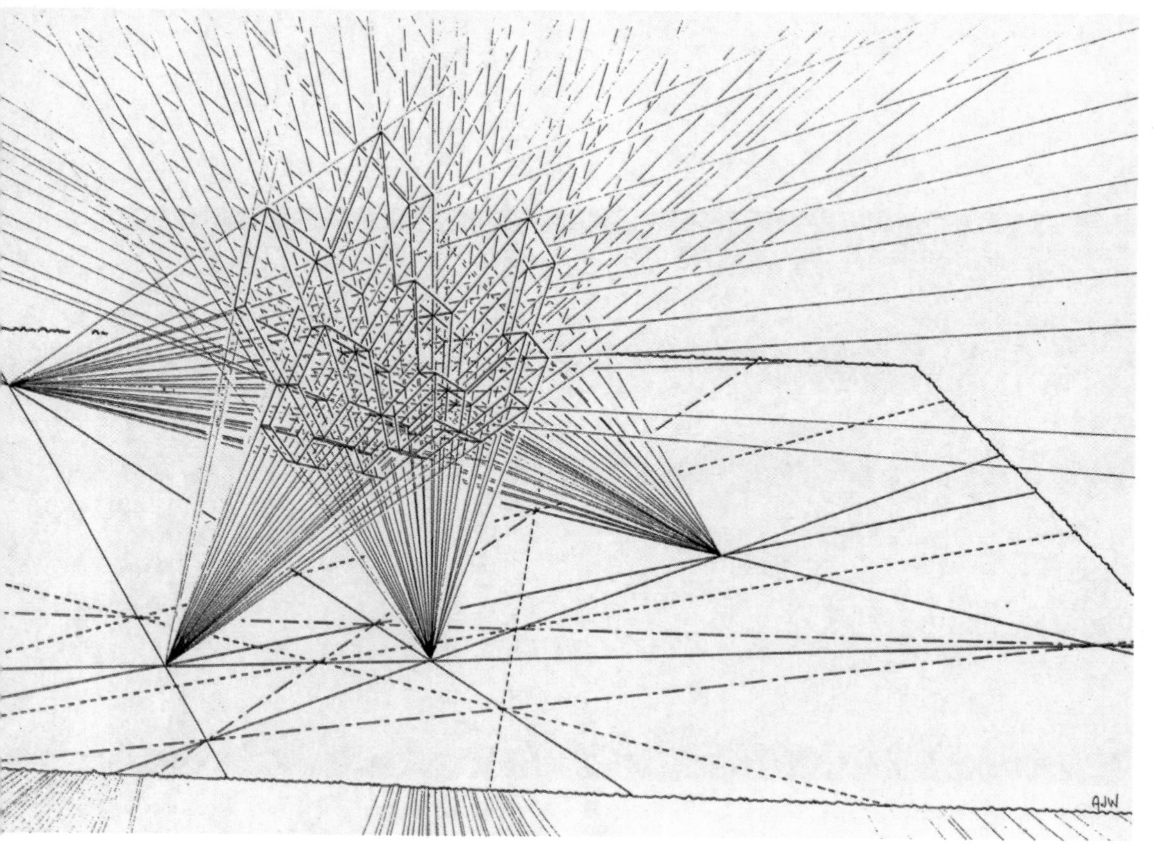

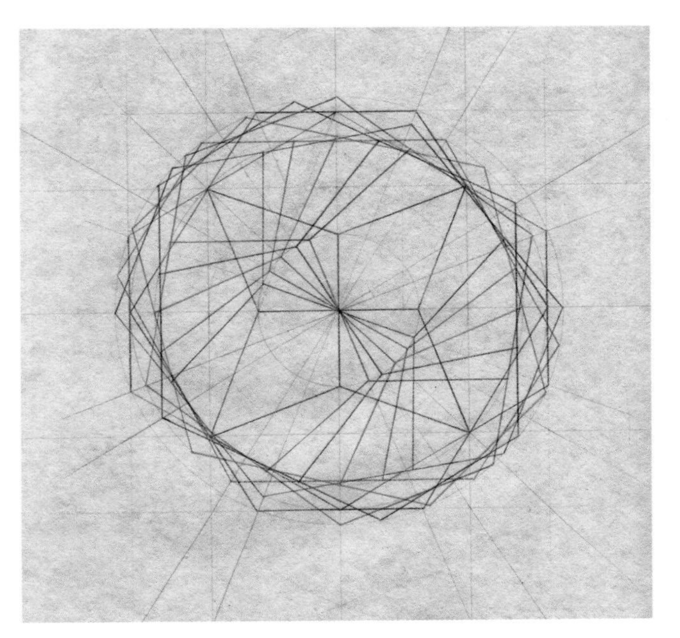

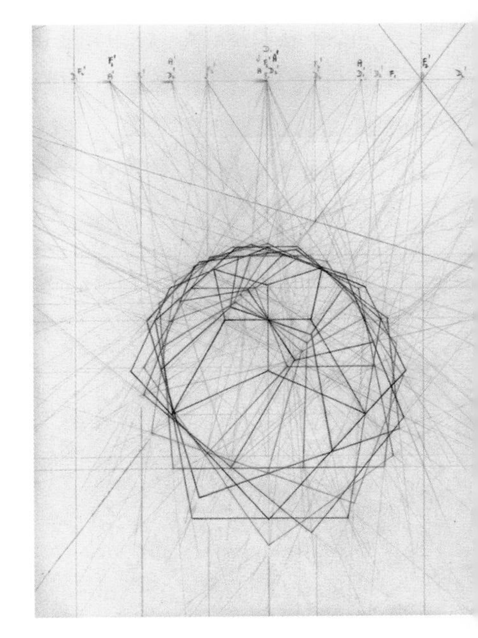

Fig. 75 *Left:* **Rotation of a regular dodecahedron section originating from a plane at infinity** *Right:* **Dodecahedral projections change within the ellipse as the points of origin move along the line of origin, brought inwards and visible at the top of the drawing**

dodecahedron. To draw this, any four lines are chosen, and generate six points (lower left). In another colour, the points that are not yet joined are connected, three further lines are necessary which in turn bisect each other at three points. Then further points can be joined which need six lines, all in a third colour; and these six lines when carefully drawn will meet in four new points. All lines should preferably be drawn across the page: 13 lines and 13 points appear at this stage.

If we move on to the dodecahedron (platonic solid composed of twelve regular pentagonal faces), which is essentially five cubes (twelve pentagrams) (Fig. 69), we have a rotational process which is obviously more complex. If we first consider a regular dodecahedron section and rotate it (Fig. 75 left), it retains its shape through six positions because, in its origin in the plane at infinity, all relevant edges (pentagon and pentagram) are parallel to each other. If we now bring that plane in from infinity and express it as a line, it contains the points from which the ellipsoidal projection is formed (Fig. 75 right). Each dodecahedral projection has changed its shape because the points of origin have moved progressively along the line of origin at the top of Fig. 75 (right).

When this line of origin is considered as the archetypal plane (edgewise), it appears in plan as a metamorphic sequence of pentagrams, the points of which, as I discovered, move on hyperbolae (Fig. 76). Having been instructed by George Adams how to draw the individual projections, I could not wait to bring them all into one drawing in order to experience how they related – something that had apparently not yet been attempted. I found the result astonishing. First of all I produced a quarter rotation of the dodecahedron (Fig. 76 left), then a half rotation of the dodecahedron (Fig. 76 right).

These drawings were produced for the first time in 1956 (Fig. 76).

From the pentagrams drawn individually in perspective it is possible to produce two-dimensional representations of three-dimensional projections of the polyhedra (see Figs 78, 79 and 80 below). Fig. 77 below shows an example of the icosahedron and within this its dual dodecahedron (blue) and finally a further icosahedron (red). This process continues within towards the point at infinity and also outwards to the plane at infinity.

These three drawings demonstrate metamorphic projections relative to point (Fig. 78), line (Fig. 79) and plane (Fig. 80). Respectively they show a threefold symmetry, a twofold symmetry (here the two symmetries are different and perpendicular to each other) and a fivefold symmetry.

These projections are intended to show a situation where the central vertical axis is perpendicular to the plane of origin. This central axis however can be chosen inclined to the plane of origin, then the three-dimensional projections will tend naturally to be asymmetrical while maintaining their special characteristics. The whole of space can be filled with such projections from each position of the archetype (plane of origin).

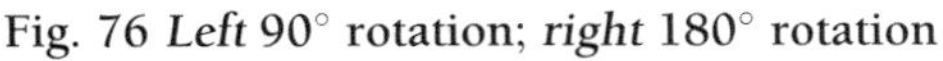

Fig. 76 *Left* 90° rotation; *right* 180° rotation

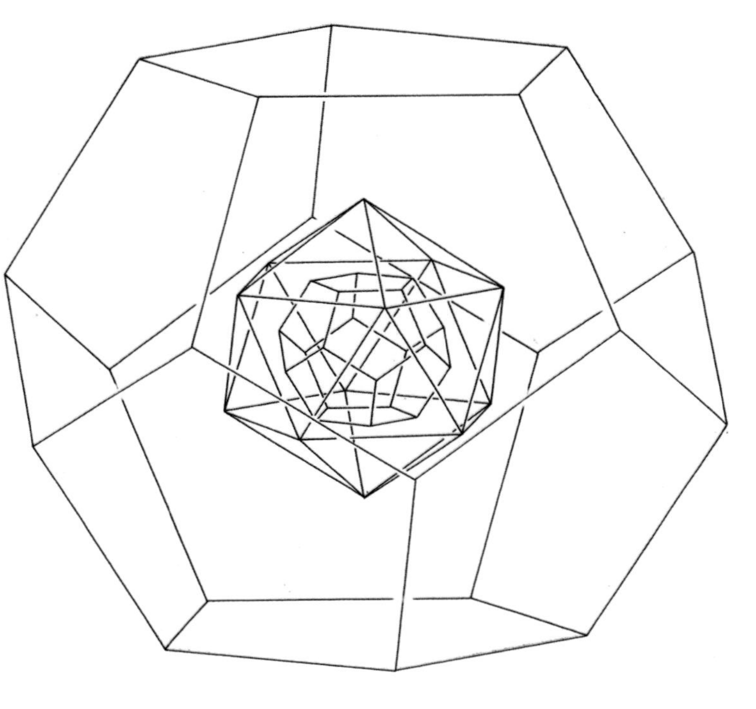

Fig. 77 *Above, left:* Outer icosahedron with inner dual dodecahedron (blue) and finally the further icosahedron (red). These metamorphic projections continue towards the point at infinity inwards and to the plane at infinity outwards
Above, right: Another way of drawing the same projection from a regular pentagram drawn in perspective below
Right: Regular projection from infinity

Fig. 78 Growing projections from a fixed archetypal plane, about the apex point, showing threefold symmetry due to the triangular orientation of the archetypal plane. This is visible as a red triangle below the projections

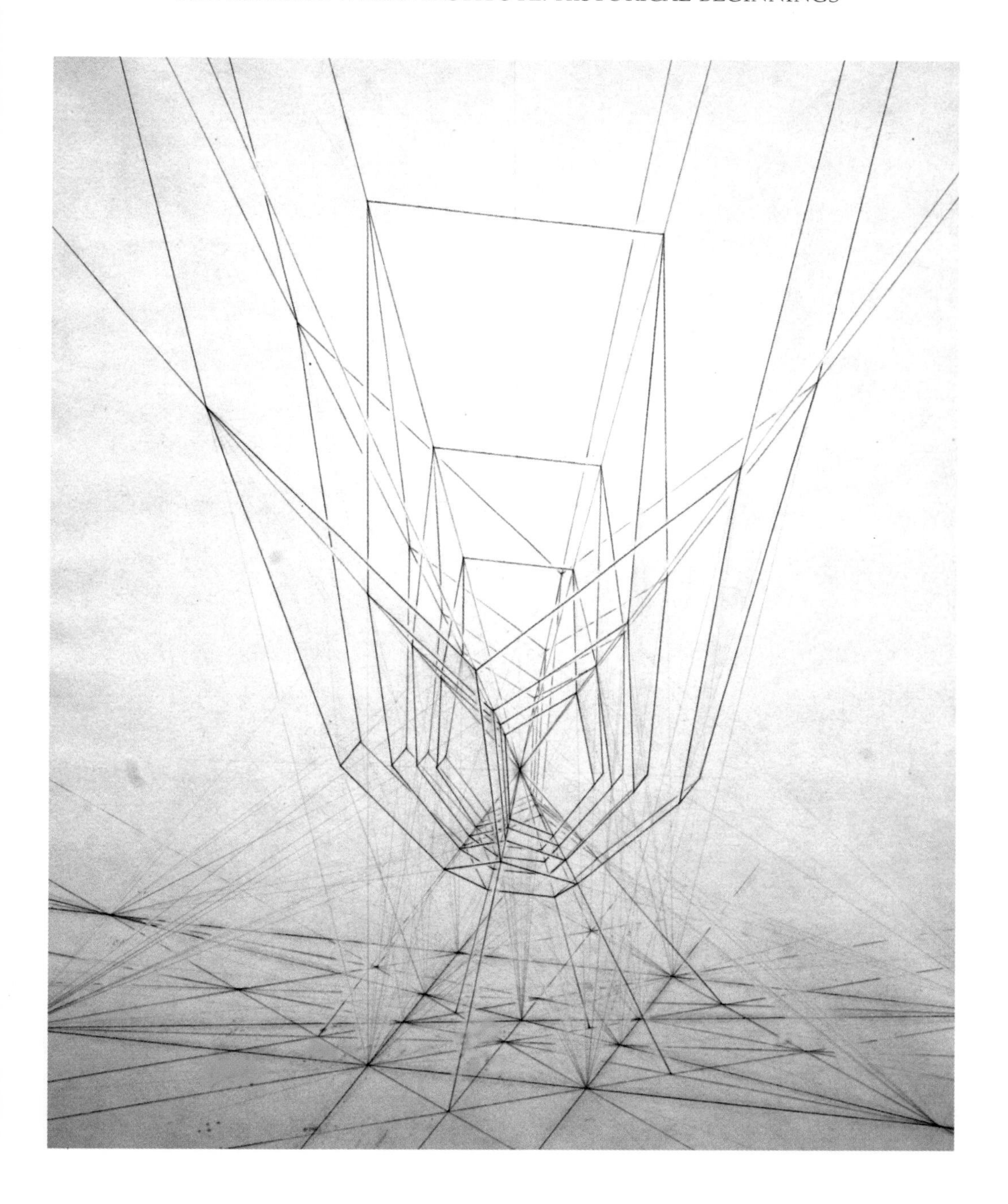

Fig. 79 Growth projections relative to the top edge showing twofold symmetry. After rotation of the system in Fig. 79, the plane of origin reaches a position where a rectangle is visible as sets of parallel red lines in the archetypal plane from which the projections emanate

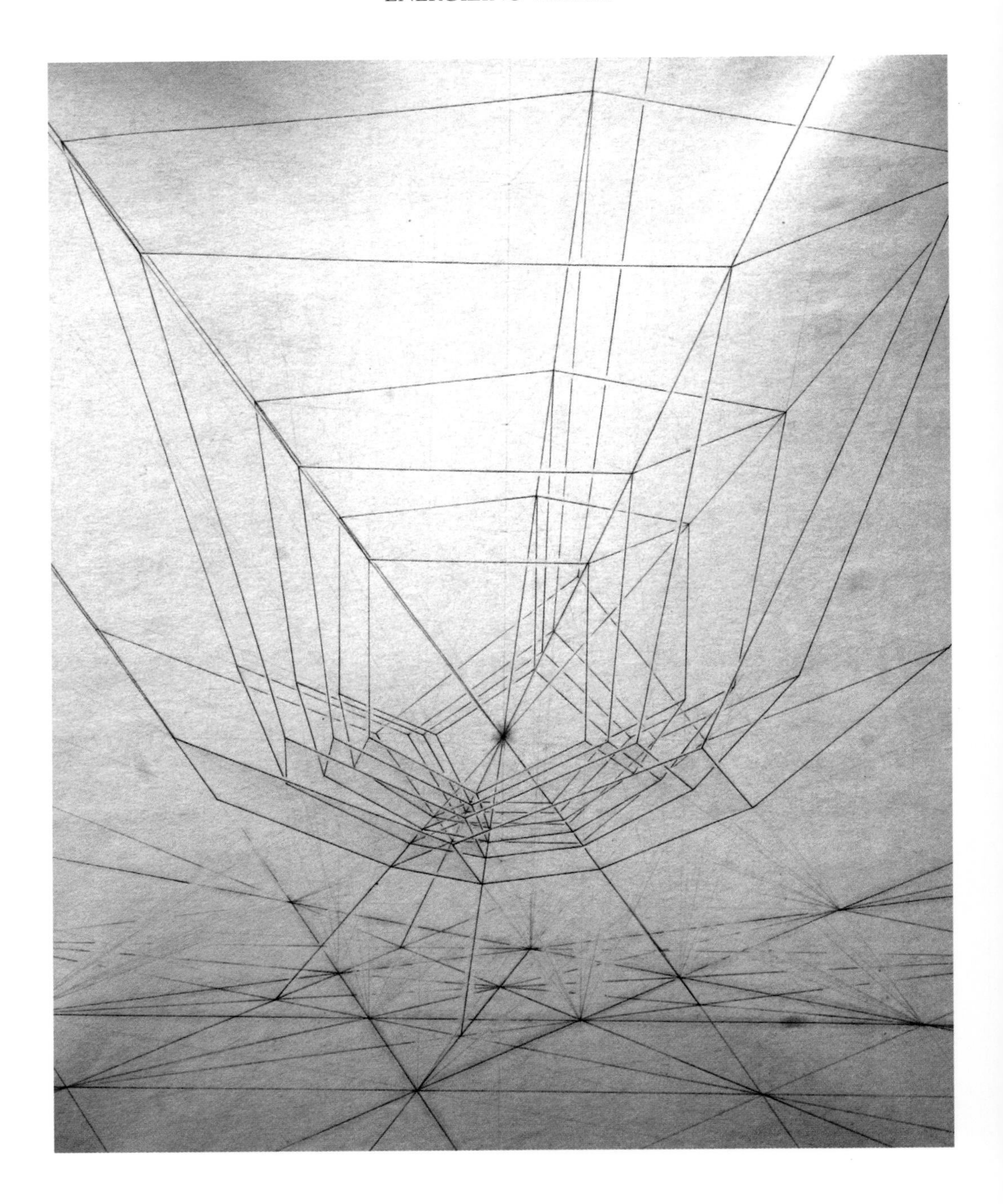

Fig. 80 Growth projections relative to a plane showing fivefold symmetry, with plane of origin in the form of a pentagram (seen in red below the projections) due to a further rotation of the system

Within each drawing, a similar process to that of the growth of a bud is demonstrated. However, to be clear, once more the three separate drawings (Figs 78, 79 and 80) demonstrate a 'rotational metamorphosis'.

The plane of origin in the last drawing is clearly a pentagram (red in Fig. 80). When the system is rotated the plane of origin reaches a position where a rectangle is visible (red in Fig. 79). On further rotation a triangular position is reached (red in Fig. 78); from these three points, the apex of the projection is, for instance, chosen.

It is clearly visible in the three separate metamorphosing projections that the three or four stages of growth in each are so designed that the last position reaches infinity. This is called a growth measure and can be designed as desired before starting. As a process it brings a specific order into the drawings.

There is very much more involved in this subject – but this would take us far beyond the scope of this chapter.

I have introduced this aspect of projective geometry here because we have discovered that when a Flowform cascade is hung inside such a structure (as shown on pages 42–3, 72–3, 85–95 ff.) the rhythmic treatment effects are altered and can be enhanced. This shows up in various test methods as illustrated above and is being further investigated.

Appendix 2
FLOW RESEARCH COLLEAGUES

Through the contemplation of ever-creating nature we make ourselves spiritually worthy of participating in her activities.

Johann Wolfgang von Goethe (1749–1832)

Adorjan, Tibor, MEng. Supplied icosidodecahedron for research

Alspach, Peter, MSc. Water quality research in New Zealand

Baxter, Mark, ARIBA. Flowform design and installation in Australia

Brown, Stuart, PhD. Reading University agriculture researcher

Burka, Uwe. Flowform use in biological purification

Charter, Simon. Project management and execution, Flowforms and biological purification

Collinson, Patrick. Waldorf class teacher and upper school science teacher

Colquhoun, Margaret, PhD. Biology, Director 'Life Science Trust' Pishwanten, Scotland

Corrin, Ian (†1992). Contributed to the casting and installation of the Akalla project 1975

Courtney, Paul, PhD. Former Brighton University lecturer, now freelance maths teacher, researcher, translator, consultant. Projective Geometry Group

Cuthbertson, David, BSc. Biophysics. Scientific advisor for the Holleman Stichting, a European organization supporting biological transmutation research. Projective Geometry Group

Damon, Betsy. Environmental art and water treatment, Flowform design in Chengdu, China

Dewdney, Rob, M.Sc. Earlier research and installation, New Zealand

Dreiseitl, Herbert. Design and research collation, water management and urban design and planning

Giorgetti, Costantino. Prof., Eng. Planning, Trieste University. Retired; trustee of the Healing Water Foundation UK

Goodwin, Pearl, Revd. Biology (embryology) lecturer

Greene, Jennifer. Early use of Flowform cascades in biological water treatment

Hecht, Chris. Development support, research collation. Trustee of the Healing Water Foundation UK

Huese, Arjen. Teacher on the biodynamic organic agriculture training at Emerson College for the last five years and is doing some research on the effect of the four etheric forces on plant forms

Hoffmann, Thomas. Hydrologist. Flowform design, planning management and execution

Jarvis, Kevin, BSc (electronic engineering). Technical director, Equipe Electronics – video displays for simulation and VR. Projective Geometry Group

Joiner, Andrew (†2008). Flowform biological purification, design, Iris Water company

Jones, Chris. Projective Geometry Group

Keis, Hanna. Flowform design production and installation. Design research

King, Paul. Treatment and testing over an extended period of time

Klingborg, Arne (†2005). Rudolf Steiner Seminariet, Järna. First major Flowform project initiator

Liess, Christian. Professor at the University of Konstanz, Germany. Retired. Advisor, now involved at the Institut für Strömungswissenschaften

Lord, Robert. Maths degree, Lasur painter, Institute colour scheme. Member of a trust interested in supporting the Institute's research

Loyter, Orit, BSc. Biology, Upper School science teacher, contributor to the biodynamic Training at Emerson College. Flowform cascade research

Malaise, Peter. Concept Manager, Ecover Belgium. Trustee of the Healing Water Foundation

Mann, Christopher. Entrepreneur. Flowform design development and research, financial support

Monzies, Michael. Flowform design and installation, France

Mueller, Peter. Wasserwerkstatt Dortmund, production and installation; supported design research

Nott, Richard. BD agriculture course, Emerson, 1987–8. BA in social psychology 1997. Projective Geometry Group

Proctor, Peter. Biodynamic research and promoter of BD Flowform installations

Provines, Ian. Prototyping design in stone and ceramics

Raab, Rex, ARIBA (†2004). Earlier project planning advisor

Raeside, Nick, MSc Forest Management. Company director, chairman BD Association UK

Sassoon, Judyth, PhD, ARCS. Biology, university research, Bern CH, Bath UK, marine biology, palaeontology. Lecturer and Flowform cascade research advisor

Schikorr, Freya. Horticulture. Early plant and biodynamic research

Schwenk, Wolfram. Institut für Strömungswissenschaften, Herrischried, Germany

Sedgman, Phil. Water treatment and effluent research, Australia

Simmonds, Neil. Teacher of tai chi. Member of Projective Geometry Group

Schwuchow, Jochen M., PhD. Engineering and Biology. Research on gravity responses in plants. HWI research volunteer collaborator

Tighe, Martin, MSc (Biochemistry)

Thomas, Nick C., MIEE, MRAeS, C.ENG. Researcher, projective geometrician, Nettlestone Laboratory, Kings Langley, Herts, UK. HWI research collaborator

Trousdell, Iain, BA, Dip Tchg, ATCL. Flowform designer, pioneer of agricultural Flowform cascade uses, Waldorf teacher, trustee of the Healing Water Foundation UK and Institute NZ

van Dijk, Paul. Flowform design prototyping and development

van Mansfeld, Jan Diek, Prof., Dr. Sc., Mondriaanlaan 67 6708, NK Wageningen, Netherlands. Warmonderhof research project.

Weidmann, Nick, BSc. Secretary Healing Water Foundation UK. Flowform design and research

Wells, Nigel. Flowform designer, agricultural and educational applications. Virbela Studio, Sweden

Wilkes, John, ARCA (Hons). Flowform concept originator. Director Flowform Design Ltd, and Healing Water Foundation UK

Wilkes, Thomas. Flowform ceramic production

Williams, Joel. Scientist at Laverstoke Park Farm. Research and promoter of Flowform cascades

Sponsors and supporters

We wish to acknowledge and thank these friends for their support over the years, especially relating to our building plans at Emerson College. It is through this support that we have been able to develop the ongoing research reported here; and this will naturally continue in future in one way or another.

Christopher Mann (UK, USA)
Kersti Biuw (Sweden)
Anthony Kaye (UK)
Katrin Ficht-Müller (Switzerland)
Unni Coward (Norway)
Susan Forsythe (UK)
Christopher Hecht (USA)
Costantino Giorgetti (UK)

For help and funding over the years, our gratitude and acknowledgment to:

Flowform Design Ltd (UK)
Emerson College (UK)
Rudolf Steiner Foundation (USA)
Software Stiftung (DL)
Rudolf Steiner Wissenschaftliche Fond (DL)
Cultura Stiftung (DL)
Mercury Arts Foundation (UK)
Margaret Wilkinson Fund (UK)
Helixor Stiftung and Fischermüehle (DL)
Cadbury Foundation (UK)
plus many individual supporters

BIBLIOGRAPHY

Adams, George, *Physical and Ethereal Spaces*, Rudolf Steiner Press, London 1965

Adams, George and Olive Whicher, *The Plant Between Sun and Earth*, Verlag Freies Geistesleben, Stuttgart 1979

Alexandersson, Olof, *Living Water*, Turnstone Press Limited, Wellingborough, Northamptonshire, 1982

Alleslev, Flemming, *Naturlig Vandbehandling*, Biologisk Projektarbejde, Århus Universitet, 1987

Baker, R., *Nature* 301, 78 (1983)

Barker, Janet, *Steigbilder, Capillary Dynamolysis, An imaging method*, Ita Wegmann Klinik, Arlesheim, Switzerland, 2005

Bartholomew, Alick, *Hidden Nature. The Startling Insights of Viktor Schauberger*, Floris Books, Edinburgh 2003

Belon, P., J. Cumps, M. Ennis, P.F. Mannaioni, M. Roberfroid, J. Sainte-Laudy, F.A. Wiegant, 'Inhibition of human basophil degranulation by successive histamine dilutions: results of a European multi-centre trial', *Inflammation Research* 48 (13):17–18 (1999)

Belon, P., J. Cumps, M. Ennis, P.F. Mannaioni, M. Roberfroid, J. Sainte-Laudy, F.A. Wiegant, 'Histamine dilutions modulate basophil activation', *Inflammation Research* 53 (5):181–8 (2004)

Benveniste, J. (1998) *From Water Memory Effects to Digital Biology*, http://www.digibio.com

Block, E.F., 'The Whole is Greater than the Sum of its Parts', *Journal of Bioelectromagnetic Medicine*, vol. 10 (2004)

Bohm, David, *Wholeness and the Implicate Order*, Routledge and Kegan Paul, London 1980

Bortoft, Henri, *The Wholeness of Nature – Goethe's Way of Science*, Floris Books, Edinburgh 1996

Briggs, John and David Peat, *Looking Glass Universe – The Emerging Science of Wholeness*, William Collins Sons & Co. Ltd, Glasgow 1984

Brückmann, Sabine, Franz-Theo Gottwald, Michael Glück, Alexandra Hoesch, Ernstfried Prade, Hans-Dirk Struwe, *Lebendiges Wasser*, Schweisfurth-Stiftung, Munich 1992

Capra, Fritjof, *The Tao of Physics*, Flamingo, Caledonian International Book Manufacturing Ltd., Glasgow 1991

Chapman, D.S., P.R. Critchlow, 'Formation of Vortex Rings from Falling Drops', *J. Fluid Mech.* 29:177–185 (1967)

Coats, Callum, *Living Energies*, Gateway Books, Bath 1996

Davenas, E., F. Beauvais, J. Arnara, M. Oberbaum, B. Robinzon, A. Miadonna, A. Tedeschi, B. Pomeranz, P. Fortner, P. Belon, J. Sainte-Laudy, B. Poitevin and J. Benveniste, 'Human basophil degranulation triggered by very dilute antiserum against IgE', *Nature* 333 (6176):816–18 (1988)

De Jonge, Gerdien B., 'Orienterend onderzoek naar de invloed van stromingsbewegingen in zgn. Wirbela Flowforms op het zelfreinigend vermogen van organisch belast slootwater', Ondersoeksverslag 1978–81, Wirbela Waterprojekt Warmonderhof, Kerk-Avezaath (1982)

Del Giudice Emilio, Giuliano Preparata 'Coherent dynamics in water as a possible explanation of biological membranes formation', *J. of Biol. Phys.* 20:105–116 (1994)

Del Giudice Emilio, Giuliano Preparata, Giuseppe Vitiello, 'Water as a Free Electric Dipole Laser', *Phys. Rev. Lett.* 61:1085–1088 (1988)

Edwards, Lawrence, *The Field of Form*, Floris Books, Edinburgh 1982

Edwards, Lawrence, *Projective Geometry*, Rudolf Steiner Institute, Phoenixville, Pennsylvania, 1985

Edwards, Lawrence, *The Vortex of Life. Nature's Patterns in Space and Time*, Floris Books, Edinburgh 1993

Emoto, Masaru, *The Messages from Water*, Hado Kyoikusha Co., Tokyo, Japan, 1999

Endres, Klaus-Peter, and Wolfgang Schad, *Moon Rhythms in Nature – How Lunar Cycles Affect Living Organisms*, Floris Books. Edinburgh 2002

Engqvist, Magda, *Gestaltkräfte des Lebendigen. Die Kupferchlorid-Kristallisation, eine Methode zur Erfassung biologischer Veränderungen pflanzlicher Substanzen*, Vittorio Klostermann, Frankfurt, Germany, 1970

Engqvist, Magda, *Die Steigbildmethode, Ein Indikator für Lebensprozesse in der Pflanze*, Vittorio Klostermann, Frankfurt, Germany, 1977

Epstein, Irving R., 'Can Droplets and Bubbles Think? *Science* 315 (5813):775–776 (2007)

Fuerstman, Michael J., Piotr Garstecki, George M. Whitesides, 'Coding/Decoding and Reversibility of Droplet Trains in Microfluidic Networks', *Science* 315 (5813):828–832 (2007)

Fyfe, Agnes, *Moon and Plant, Capillary Dynamic Studies*, Society for Cancer Research, Arlesheim, Switzerland, 1967

Geiger Wolfgang, Herbert Dreiseitl, *Neue Wege für das Regenwasser, Handbuch zum Rückhalt und zur Versickerung von Regenwasser in Baugebieten*, R. Oldenbourg Verlag GmbH, Münich, Germany, 1995

Goethe, Johann Wolfgang von, *The Metamorphosis of Plants*, Bio-Dynamic Farming and Gardening Association, New York 1993

Gross, Peter (2000a) *Geniales Wunderwesen Wasser*. http://www.wasser-informationen.de/wasserinformationen/erfahrungen/index.html

Gross, Peter (2000b) *Wasser*, http://www.wasserinformationen.de/wasserinformationen/erfahrungen/index.html

Hagel, Ingo, *Unbehandeltes, Flowform-behandeltes und gepumptes Wasser im Pflanzenversuch*, Emerson College, Forest Row, England, 1983

Hatschek, Emil, A Study of the Forms Assumed by Drops and Vortices of a Gelatinizing Liquid in Various Coagulating Solutions', *Proc. Roy. Soc. Ser. A*, vol. 95, no. 669:303–316 (1919)

Hoesch, Alexandra, et al., *Lebendiges Wasser* [Living Water], Schweisfurth-Stiftung, Munich, Germany, 1992

Jahnke, Dittmar, *Sensibles Wasser Nr. 2, Morphologische Typisierung von Tropfenbildversuchen und Tropfenbildern, und Morphologische Unterscheidungs-merkmale für die Auswertung von Wasserqualitäts-Untersuchungen mit der Tropfenbildmethode, Verein für Bewegungsforschung* e.V., Institut für Strömungswissenschaften, Herrischried, Germany, 1993

Kilner, Philip, 'The Single Cavity Flowform', *The Goetheanum*, vol. 5, no. 1, Dornach, Switzerland

Kolisko, Eugen and Lily, *Agriculture of Tomorrow*, Kolisko Archive Publishing, Bournemouth, England, 1978

Kristiansen, U.R., H.E. Berntsen, T. Lunde, C. Thaulow, A. Rørdal, *Sound Measurements on a Sevenfold Flowform*, The Water Research Group in Trondheim, Norwegian Hydrodynamic Laboratories, University of Trondheim, Norway, 1993

Kröplin, Bernd (ed.), *Welt im Tropfen – Gedächtnis- und Gedankenformen im Wasser*, Gutesbuchverlag ISD, Universität Stuttgart, Germany, 2001

Lauterwasser, Alexander, *Wasser Klang Bilder*, AT Verlag. Aarau-München, Germany, 2003

Ludwig, Wolfgang, 'Alle Lebensprozesse sind

unmittelbar mit dem Wasser verbunden. Daher kommt dem Wasser in der Umweltproblematik eine besondere Stellung zu', in Talkenberger, Peter P., Treven, Michael, *Umweltmedizin – Ein neues Zeitalter der Gesundheit*, Möwe Verlag, Idstein 1999, pp. 71–9

Marrin, West, *Universal Water. The Ancient Wisdom and Scientific Theory of Water*, Inner Ocean Publishing Inc., Maui, Hawaii, 2002

Mæhlum, Trond *Økologisk Avløpsrensing, 1. Bruk av konstruerte våtmarker for rensing av avløpsvann i Norge, 2. Effekten av strømningsformer (Flowforms) for lufting biodammer om vinteren, 3. Wastewater treatment by constructed wetlands in Norwegian climate: Pretreatment and optimal design*, Jordforsk, Norges Landbrukshøgskole Ås, 1991

Nickel, Erwin, 'Die Reproduzierbarkeit der sogenannten "empfindlichen Kupferchloridkristallisation"', *Bulletin der Naturforschenden Gesellschaft Freiburg, Bull. Soc. Frib. Sc. Nat.*, vol. 57, Fasc. II, Universitätsverlag Freiburg, Switzerland, 1968

Paul, J.B., C.P. Collier, R.J. Saykally, 'Direct Measurement of Water Cluster Concentrations by Infrared Cavity Ringdown Laser Absorption Spectroscopy', *J. Phys. Chem. A* 101:5211–5214 (1997)

Pearsaal, Joanna, and Bryan Innes 'Natural Swimming Pool', *Natural in-formation*, Newsletter for the Permaculture Institute of NZ, Auckland, New Zealand, 2000

Peitgen, Heinz-Otto, Hartmut Jürgens, Dietmar Saupe, *Fractals for the Classroom, Part 1: Introduction to Fractals and Chaos*, Springer-Verlag, Berlin-Heidelberg-New York-Tokyo **1992a**

Peitgen, Heinz-Otto, Hartmut Jürgens, Dietmar Saupe, *Fractals for the Classroom, Part 2: Complex Systems and Mandelbrot Set*, Springer-Verlag, Berlin-Heidelberg-New York-Tokyo **1992b**

Peitgen, Heinz-Otto, and Peter H. Richter, *The Beauty of Fractals: images of complex dynamical systems*, Springer-Verlag, Berlin-Heidelberg-New York-Tokyo, **1986**

Peitgen, Heinz-Otto, and Dietmar Saupe (eds), *The Science of Fractal Images*, Springer-Verlag, Berlin-Heidelberg-New York-Tokyo **1988**

Pfeiffer, Ehrenfried, *Studium von Formkräften an Kristallisationen*, Dornach, Switzerland, 1931

Pfeiffer, Ehrenfried, *Empfindliche Kristallisationen als Nachweis von Formkräften im Blut*, Emil Weises Verlag, Dresden 1935

Pollack, Gerald H., *Cells, Gels and the Engines of Life*, Ebner and Sons, Seattle 2001

Prakash, Manu and Neil Gershenfeld, 'Microfluidic Bubble Logic', *Science* 315 (5813):832 (2007)

Presti, D., Pettgrew, J., *Nature* 285, 99 (1980)

Prigogine, Ilya, *From Being to Becoming: Time and Complexity in the Physical Sciences*, W.H. Freeman and Co., San Francisco 1980

Proctor, Peter, *Biodynamics, A New Direction in Farming and Gardening in New Zealand*, Random House, 1989

Proctor, Peter, *Grasp the Nettle*, Random House, 1997

Rasmus, Jansson, *Dowsing: Science or Humbug?* (1998; revised January 1999)

Rocard, Y., *La Recherche* 12, 792 (1981)

Rogers, W.B., 'On the Formation of Rotating Rings by Air and Liquids under Certain Conditions of Discharge', *Am. J. Sci. Arts. Sec. Ser.* 26:246–58 (1858)

Schauberger, Viktor, 'Implosion statt Explosion', in *Implosion. Biotechnische Nachrichten*, Verein fur Implosionsforschung und Anwendung e.V., Zell a.H., Germany, September 2006, vol. 153:7–20 (1955)

Schauberger, Viktor, *Nature as Teacher – New Principles in the Working of Nature*, Gateway Books, Bath, 1998a

Schauberger, Viktor, *The Water Wizard – The Extraordinary Properties of Natural Water*, Gateway Books, Bath 1998b

Schauberger, Viktor, *The Energy Evolution –*

Harnessing Free Energy from Nature, Gateway, Gill & Macmillan Ltd, Dublin 2000

Schiff, Michel, *The Memory of Water*, Thorsons, London 1995

Schikorr, Freya, 'Wheat Germination Responses to Flowform Treated Water', *Star and Furrow* 74:15–22 (1990)

Schikorr, Freya, 'A Comparison of Different Methods of Stirring the Biodynamic Field Preparations', *Star and Furrow* 82:12–16 (1994)

Schmidt, Georg W., *Aufbau lebensfähiger Naturbereiche als Gestaltungsaufgabe in bedrohten oder zerstörten Landschaften, Lebendige Erde*, Nr. 2, Darmstadt (1984), pp. 54–65

Schneider, Michael S., *Constructing the Universe – The Mathematical Archetypes of Nature, Art and Science*, Harper Perennial, New York 1995

Schönberger Christian, Christian Liess, *Wirksamkeit der Flowforms, Zusammenstellung und Auswertung der bis 1994 durchgeführten Untersuchungen über die Wirkung der Virbela-Flowforms*, Atelier Dreiseitl, Überlingen, Germany, 1995

Schwenk, Theodor, *Bewegungsformen des Wassers*, Verlag Freies Geistesleben, Stuttgart, Germany, 1967

Schwenk, Theodor, Water – *The Element of Life*, Anthroposophic Press, Bell's Pond, Hudson, New York, 1989

Schwenk, Theodor, *Sensitive Chaos. The Creation of Flowing Forms in Water and Air*, revised 2nd ed. Rudolf Steiner Press, London, 1996. (First edition: *Das Sensible Chaos*, Verlag Freies Geistesleben, Germany, 1962)

Schwenk, Wolfram (ed.) *Sensibles Wasser Nr. 6, Schritte zur positiven Charakterisierung der Wassers als Lebensvermittler*, Verein für Bewegungsforschung e.V., Institut für Strömungswissenschaften, Herrischried, Germany, 2001

Seamon, David and Arthur Zajonc (eds) *Goethe's Way of Science: A Phenomenology of Nature*, State University of New York Press, 1998.

Selawry, A. and O., *Die Kupferchlorid-Kristallisation in Naturwissenschaft und Medizin*, Gustav Fischer Verlag, Stuttgart, Germany, 1957

Sernbo Krister, Lars Fredlund, *Avloppsbehandling i biologiska dammar*, Järna 1991

Sheldrake, Rupert, *The Presence of the Past*, Hartnolls Limited, Bodmin, England, 1988

Smith, C.W., 'Electromagnetic Effects in Humans', in *Biological Coherence and Response to External Stimuli*, Springer-Verlag, Berlin 1987

Smith, Howard J., 'A Study of Some of the Parameters Involved in the Drop Picture Method', *Bericht* 111/1974, Max-Plank-Institut für Strömungsforschung, Göttingen 1974

Smith, Howard J., 'The Hydrodynamic and Physico-chemical Basis of the Drop Picture Method', *Bericht* 8/1975, Max-Plank-Institut für Strömungsforschung, Göttingen 1975

Spencer, Tod, *A Study of the Effectiveness of Flowforms and their Suitability for Greywater Recycling.* Integrated Project in Conservation Technology, Southern Cross University, Australia, 1995

Steiner, Rudolf, *The Agricultural Course*, Koberwitz, Silesia, 7–16 June 1924, Rudolf Steiner Publishing Co., London 1938

Steiner, Rudolf, *Spiritual Science and Medicine*, lectures in Dornach from 21 March to 9 April 1920, Rudolf Steiner Publishing Co., London 1975

Steiner, Rudolf, *Nature's Open Secret: Introductions to Goethe's Scientific Writings*, Rudolf Steiner Press, London 2000

Strid, Martin, *Rhythmisk Strömning – En Studie av Virbela Flödesformer*, Tekniska Högskolan i Luleå, Examensarbete, Luleå, Sweden, 1984

Strube, Jürgen, and Peter, Stolz, *Verbesserte Wasserqualität zum Brotbacken durch simulierten Bergbach mit Flowformen* [Improved quality of water for baking bread] Dipperz, Kwalis Qualitätsforschung, Fulda 1999

Szent-Gyorgyi, A., 'To See What No One Has Thought', *Biological Bulletin* 175:191–241 (1988)

Tebbut, T.H.Y., *Principles of Water Quality Control*, Pergamon Press, Oxford 1992

Thomas, Nick C., 'Flowform Rhythms', *Science Forum* No. 4: 16–19 (1983). Science Group of the Anthroposophical Society in Great Britain.

Thompson, D'Arcy, *On Growth and Form*, Cambridge University Press, Cambridge 1942

Thomson, J.J., H.F. Newall, 'On the Formation of Vortex Rings by Drops Falling into Liquids, and some Allied Phenomena', *Proc. Roy. Soc.* 39:417 (1885)

Thun, Maria, *The Biodynamic Sowing and Planting Calendar*, Floris Books, Edinburgh 2003

Thun, Maria and Matthias, *The Biodynamic Sowing and Planting Calendar*, Floris Books, Edinburgh 2004

Thun, Maria and Matthias, *The Biodynamic Sowing and Planting Calendar*, Floris Books, Edinburgh 2006

Tingstad, Annette, *Quality and Method. Rising Pictures in Evaluation of Food Quality*, Gads Forlag, Copenhagen 2002

Trousdell, Iain, *Virbela Flowforms & Biodynamic Preparations Stirring*, Hastings, NZ, 1991

Trousdell, Iain, 'Flowform Design Types and Oxygenation', Healing Water Institute NZ *Proceedings*, Hastings, NZ, 1989

Van Mansfelt, J.D., *To Whom it May Concern*, Warmonderhof, Kerk-Avezaath, Holland, 1986

Wagenaar, Walter, 'Beweging: Sturing In Levensprocessen van het Water? Wirbela Waterprojekt Warmonderhof', Ondersoeksverslag, Kerk-Avezaath 1984

Wilkens, Andreas, Michael Jacobi, Wolfram Schwenk, *Understanding Water*, Floris Books, Edinburgh 2005

Wilkes, John, 'Flow Design Research Relating to Flowforms', in 'Chaos, Rhythm and Flow in Nature', *The Golden Blade* 46, Floris Books, Edinburgh, 1993

Wilkes, John, 'Water as a Mediator for Life', in *Elemente der Naturwissenschaft*, Naturwissenschaftliche Sektion am Goetheanum, Dornach, Switzerland, vol. 74:34–51 (2001)

Wilkes, John, *Flowforms: The Rhythmic Power of Water*, Floris Books, Edinburgh 2003 (2nd printing 2005, 3rd printing 2009). (*Flowforms. O poder ritmico da água*, Senac, São Paulo 2008; *Das Flowform Phänomen. Die verborgene rhythmische Energie des Wassers*, Engel Verlag, Stuttgart 2008; *Il Fenomeno Flowform – l'energia ritmica latente dell'acqua*, editrice antroposofica, Milan 2010)

Worrall, Peter, 'The Reed Bed Revolution', *Landscape Design* 208: 16–18 (1992)